Lecture Notes in Computer Science 2409

Edited by G. Goos, J. Hartmanis, and J. van Leeuwen

Springer
Berlin
Heidelberg
New York
Barcelona
Hong Kong
London
Milan
Paris
Tokyo

David M. Mount   Clifford Stein (Eds.)

Algorithm Engineering and Experiments

4th International Workshop, ALENEX 2002
San Francisco, CA, USA, January 4-5, 2002
Revised Papers

 Springer

Series Editors

Gerhard Goos, Karlsruhe University, Germany
Juris Hartmanis, Cornell University, NY, USA
Jan van Leeuwen, Utrecht University, The Netherlands

Volume Editors

David M. Mount
University of Maryland, Department of Computer Science
College Park, MD 20742, USA
E-mail: mount@cs.umd.edu

Clifford Stein
Columbia University, Department of IEOR
500W. 120 St., MC 4704
New York, NY 10027, USA
E-mail: cliff@ieor.columbia.edu

Cataloging-in-Publication Data applied for

Die Deutsche Bibliothek - CIP-Einheitsaufnahme

Algorithm engineering and experimentation : 4th international workshop ;
revised papers / ALENEX 2002, San Francisco, CA, USA, January 4 - 5, 2002.
David M. Mount ; Clifford Stein (ed.). - Berlin ; Heidelberg ; New York ;
Barcelona ; Hong Kong ; London ; Milan ; Paris ; Tokyo : Springer, 2002
   (Lecture notes in computer science ; Vol. 2409)
   ISBN 3-540-43977-3

CR Subject Classification (1998): F.2, E.1, I.3.5, G.2

ISSN 0302-9743
ISBN 3-540-43977-3 Springer-Verlag Berlin Heidelberg New York

This work is subject to copyright. All rights are reserved, whether the whole or part of the material is
concerned, specifically the rights of translation, reprinting, re-use of illustrations, recitation, broadcasting,
reproduction on microfilms or in any other way, and storage in data banks. Duplication of this publication
or parts thereof is permitted only under the provisions of the German Copyright Law of September 9, 1965,
in its current version, and permission for use must always be obtained from Springer-Verlag. Violations are
liable for prosecution under the German Copyright Law.

Springer-Verlag Berlin Heidelberg New York,
a member of BertelsmannSpringer Science+Business Media GmbH

http://www.springer.de

© Springer-Verlag Berlin Heidelberg 2002
Printed in Germany

Typesetting: Camera-ready by author, data conversion by PTP-Berlin, Stefan Sossna e.K.
Printed on acid-free paper   SPIN: 10873691   06/3142   5 4 3 2 1 0

Preface

The annual workshop on Algorithm Engineering and Experiments (ALENEX) provides a forum for the presentation of original research in the implementation and experimental evaluation of algorithms and data structures. ALENEX 2002 was the fourth workshop in this series. It was held in San Francisco, California on January 4–5, 2002. This volume collects extended versions of the 15 papers that were selected for presentation from a pool of 34 submissions.

We would like to thank the sponsors, authors, and reviewers who helped make ALENEX 2002 a success. We also want to thank the invited speakers, Cynthia Phillips of Sandia National Laboratories, Martin Farach-Colton of Google, and Michael Kass of Pixar. Finally, we would like to thank Springer-Verlag for publishing these papers in their *Lecture Notes in Computer Science* series.

May 2002
David M. Mount
Clifford Stein

ALENEX 2002 Sponsors

The following organizations provided direct financial support, which enabled us to host invited speakers and provide reduced registration fees for students.
- Sandia National Laboratories
- Akami Technologies Inc.
- NEC Research

The following provided in-kind support, facilitating the workshop.
- SIAM, the Society for Industrial and Applied Mathematics
- SIGACT, the ACM SIG on Algorithms and Computation Theory
- Columbia University

ALENEX 2002 Program Committee

Nancy Amato (Texas A&M University)
Marshall Bern (Xerox PARC)
Michael Goodrich (University of California, Irvine)
Tom McCormick (University of British Columbia)
Michael Mitzenmacher (Harvard University)
David Mount (University of Maryland; Co-chair)
Giri Narasimhan (Florida International University)
Rajeev Raman (University of Leicester)
Clifford Stein (Columbia University; Co-chair)

ALENEX 2002 Steering Committee

Michael Goodrich (University of California, Irvine)
Adam Buchsbaum (AT&T Labs)
Roberto Battiti (University of Trento, Italy)
Andrew V. Goldberg (Intertrust STAR Lab)
Michael T. Goodrich (University of California, Irvine)
David S. Johnson (AT&T Bell Laboratories)
Catherine C. McGeoch (Amherst College)
Bernard M.E. Moret (University of New Mexico; chair)
Jack Snoeyink (UNC-Chapel Hill)

Table of Contents

ALENEX 2002

On the Implementation of MST-Based Heuristics for the Steiner Problem in Graphs

Marcus Poggi de Aragão and Renato F. Werneck*

Department of Informatics, Catholic University of Rio de Janeiro, R. Marquês de São
Vicente, 225, Rio de Janeiro, RJ, 22453-900, Brazil.
poggi@inf.puc-rio.br, rwerneck@cs.princeton.edu

Abstract. Some of the most widely used constructive heuristics for the
Steiner Problem in Graphs are based on algorithms for the Minimum
Spanning Tree problem. In this paper, we examine efficient implementations of heuristics based on the classic algorithms by Prim, Kruskal,
and Borůvka. An extensive experimental study indicates that the theoretical worst-case complexity of the algorithms give little information
about their behavior in practice. Careful implementation improves average computation times not only significantly, but asymptotically. Running times for our implementations are within a small constant factor
from that of Prim's algorithm for the Minimum Spanning Tree problem,
suggesting that there is little room for improvement.

1 Introduction

The Steiner Problem in Graphs (SPG) can be stated as follows: given an undirected graph $G = (V, E)$, a set of terminals $T \subseteq V$, and a positive cost function
$c(v, w)$ for all $(v, w) \in E$, find a minimum weight connected subgraph of G containing all vertices in T. This is one of the most studied NP-hard problems. A
wealth of links to recent papers on this subject can be found in [5].

In this paper, we are especially interested in *constructive heuristics*, algorithms that build feasible solutions from scratch. They can be used as stand-alone
algorithms, providing solutions of reasonable quality within a short amount of
time. However, their application is much broader. They are often used as sub-
routines of more elaborate primal heuristics [2,10], dual heuristics [8], and exact
algorithms [4,8]. Regardless of the application, constructive heuristics should be
as fast as possible, while preserving solution quality.

A number of such heuristics are described in the literature (see [3,12] for
surveys). We focus our attention on heuristics that are direct extensions of exact
algorithms for the Minimum Spanning Tree (MST) problem. The main contribution of this paper are new, fast implementations for these methods. This includes
not only versions of the traditional methods based on Prim's and Kruskal's algorithms, but also a Borůvka-based heuristic, a natural extension that has not

---

* Current address: Department of Computer Science, Princeton University, Princeton,
NJ, 08544.

D. Mount and C. Stein (Eds.): ALENEX 2002, LNCS 2409, pp. 1–15, 2002.
© Springer-Verlag Berlin Heidelberg 2002

been previously described in the literature. (Descriptions of the original MST algorithms can be found in [1].) Extensive computational experiments suggest that these implementations allow the heuristics to run within a constant factor of "pure" MST algorithms on average, even though they have higher theoretical worst-case complexities. We show that this is not the case when some of the previous implementations are used.

The remainder of this paper is organized as follows. Section 2 presents the Distance Network Heuristic with its implementations and variations, alongside with a description of the Voronoi diagram for graphs, a data structure that is central in this text. Implementations of heuristics based on Borůvka's, Kruskal's, and Prim's algorithms are described in Sections 3, 4 and 5, respectively. An empirical comparative analysis is presented in Section 6. Final remarks are made in Section 7.

2   Distance Network Heuristic (DNH)

Given a weighted graph $G = (V, E)$, we define the corresponding *distance network* $D_G = (V, E')$ as the graph containing all vertices in V and such that for all pairs $(v, w) \in V$, there is an edge in D_G with cost equal to the length of the shortest path between v and w in G. The Distance Network Heuristic (DNH) consists of the following three steps:

(a) find the MST of the subgraph of D_G induced by T, then expand its edges (with the associated shortest paths) to create a valid Steiner tree of G;
(b) find the MST of the subgraph of G induced by the solution obtained in (a);
(c) eliminate all non-terminal vertices with degree one.

Since step (a) already generates a valid solution, steps (b) and (c) can be seen as a post-optimization procedure; we call it MST-prune, following [2]. Step (b) takes at most $O(|E| + |V| \log |V|)$ time[1] and step (c) requires $O(|V|)$ time.

The best known implementation of step (a) is due to Melhorn [6]. Essential to his implementation of DNH — and to some of the implementations proposed in this paper for other heuristics — is the concept of *Voronoi diagrams* in graphs. Given a set $X \subseteq V$ of vertices, the Voronoi diagram is a partition of V into *Voronoi regions*, one for each vertex in X. The Voronoi region of x_i (denoted by $vor(x_i)$) is defined as the set of vertices in V that are not closer to any other vertex in X. Ties are broken arbitrarily. If $v \in vor(x_i)$, x_i is said to be the *base* of v, denoted by $x_i = base(v)$.

Let $p(base(v), v)$ denote the shortest path from $base(v)$ to v, let $d(v)$ be the length of this path, and let $pred(v)$ be the vertex that immediately precedes v in the path. The Voronoi diagram with respect to T, defined by the values $base(v)$,

---

[1] In the theoretical part of this paper, whenever we need a "pure" algorithm for calculating minimum spanning trees, we use Prim's algorithm, implemented with Fibonacci heaps. We also assume this data structure is used in implementations of Dijkstra's algorithm for shortest paths. See [1] for alternative implementations.

$d(v)$ and $pred(v)$ for all $v \in V$, can be computed in $O(|E| + |V| \log |V|)$ time with Dijkstra's algorithm (taking all terminal nodes as sources).

An edge (v, w) will be considered a *frontier edge* if $base(v) \neq base(w)$. Let G' be a graph with the same set of edges and vertices as G, but edge costs given by cost function c', defined as follows. If (v, w) is a frontier edge, then $c'(v, w) = d(v) + c(v, w) + d(w)$, else $c'(v, w) = 0$. Note that every frontier edge (v, w) is uniquely associated by this cost function to a path between $base(v)$ and $base(w)$. Melhorn [6] shows that there exists an MST of D_G containing only paths associated to edges with positive cost (i.e., frontier edges) in the MST of G'. There is no need to actually compute D_G.

Therefore, step (a) of Melhorn's implementation of DNH is actually divided into two substeps: (i) calculating the Voronoi diagram of G; and (ii) finding an MST of G' and extracting the corresponding $|T| - 1$ paths in G. We have seen that substep (i) requires $O(|E| + |V| \log |V|)$ operations. We now consider different implementations of substep (ii).

DNH-Prim. If Prim's algorithm is used to determine the MST of G', as suggested by Melhorn, DNH will require $O(|E| + |V| \log |V|)$ operations. We refer to this implementation as DNH-Prim.

DNH-Borůvka. Alternatively, a slightly modified version of Borůvka's algorithm (also known as Sollin's algorithm) can be used in substep (ii). The classic algorithm starts with $|V|$ components, one associated to each vertex in V. It proceeds by repeatedly performing a step in which every component is connected (by a single edge) to the closest neighboring component. Because the number of components is at least halved in each step, there will be only one component left after at most $\log |V|$ steps. Each step can be performed in $O(|E|)$ time, thus leading to a $O(|E| \log |V|)$ algorithm.[2]

Our "modified version" considers only the vertices in T as initial components. In each of the connecting steps, every component is linked to its closest counterpart by a path with one or more edges. Finding all links in a given step amounts to checking the nonzero-cost edges in G' and using the Voronoi diagram to determine the associated paths in G. This can be done in $O(|E|)$ time for all paths in a given step. The Steiner tree can therefore be found in $O(|E| \log |T|)$ time, increasing the algorithm's complexity to $O(|E| \log |T| + |V| \log |V|)$. We note that T may be a small set when compared to V, so this alternative implementation, called DNH-Borůvka, is often faster than DNH-Prim.

DNHz. A slight modification in DNH-Borůvka can improve average solution quality while preserving computational complexity. Whenever a non-terminal vertex v is added to the solution in a connecting step, we can set $d(v)$ (its distance to the closest terminal) to zero. After all, v is now part of a component, as all terminals are, so it can be used to link different components. Since edge

---

[2] This can actually be implemented in $O(|E| \log \log |V|)$ time, but our algorithm for the SPG is based on the less efficient basic implementation.

costs in G' are computed dynamically, some will be reduced by this operation. This allows for future connecting steps to take into account (to a certain extent) the fact that there is already a partial solution, leading to final solutions of better quality. We call DNHz this modification of DNH-Borůvka.

3   Borůvka

This section describes a heuristic to the SPG based on Borůvka's algorithm for the MST problem. To the best of our knowledge, there has been no previous description of such heuristic in the literature. However, this is just a natural extension of an existing idea, adapting MST algorithms to solve the Steiner problem in graphs. When compared to Prim's and Kruskal's heuristics, it turns out that Boruvka's has smaller worst-case running times, but tends to produce solutions of slightly worse quality, as Section 6 shows.

The heuristic starts with the set T of terminals as its initial set of components. Then, in each iteration every component in the partial solution is linked to the closest component by the shortest possible path. Every such link creates a new component, formed by the original vertices and those in the shortest path. After at most $\log |T|$ iterations, only one component (a Steiner tree) will be left.

To implement this algorithm efficiently, we keep a Voronoi diagram, updated before each step to include (as bases) new vertices added to the solution. With the diagram, we can find the closest counterpart of each component in $O(|E|)$ time (for all components) by checking all edges (v, w) such that $base(v)$ and $base(w)$ belong to different components (this is a subset of the frontier edges).

Building the first Voronoi diagram requires a complete execution of Dijkstra's algorithm, having all terminal nodes as sources. In each of the following iterations, the Voronoi diagram is just *updated*: Dijkstra's algorithm is run having as sources just the vertices added to the solution in the previous iteration (with zero distance label, since they become potencial bases). Because distance labels of previous diagrams are still valid upper bounds, updating can be significantly faster than building a diagram from scratch: they tend to be limited to a few local operations. In the worst case, however, the update will require $O(|E| + |V| \log |V|)$ operations. The overall complexity of Borůvka's heuristic, therefore, is $O((\log |T|)(|E| + |V| \log |V|))$.

Note that Borůvka uses essentially the same strategy as DNHz to improve solution quality, but takes it one step further. In DNHz, every vertex added to the solution has its distance label set to zero. This helps the algorithm find *shortcuts* — paths whose ends are not both terminals — between components in future iterations. In Borůvka's heuristic, the entire Voronoi diagram is updated; not only do vertices already in the solution have their distances set to zero, but also their neighbors may have their distances reduced. Shortcuts are more likely to be found, thus increasing average solution quality. Although there is an extra $\log |T|$ factor in the algorithm's worst-case performance, Section 6 shows that in practice Borůvka is only marginally slower than DNH-Borůvka or DNHz. This confirms that updating the Voronoi diagram is usually a "cheap" operation.

4   Kruskal

Kruskal's algorithm for the MST problem can also be used as a starting point for a constructive heuristic for the SPG. The classic algorithm starts with a partial solution with all $|V|$ vertices and no edges. Then, it adds edges to the solution sequentially, always selecting the shortest edge that does not create a cycle. The algorithm ends after $|V| - 1$ edges are selected, when only one component will be left. Wang [13] suggested a Kruskal-based heuristic for the SPG. In his algorithm, the initial partial solution is made up only by the set of terminals (T). The algorithm then sequentially adds *paths* to the solution (including intermediate non-terminal vertices), always selecting the shortest path connecting different components. After $|T| - 1$ such paths are added, a valid solution is found.

4.1   Basic Implementation (Kruskal-B)

Kruskal's heuristic can be implemented as follows. Let *dist* be a structure that maps each connected component to its closest counterpart (and contains the distance between them). In the beginning, each component is made up by a single terminal. The initialization of *dist* requires $O(|T|)$ calls to Dijkstra's algorithm, each having one terminal as source and any other terminal as target. In each of the $|T| - 1$ iterations that follow, we (i) scan *dist* to determine the closest pair of components; (ii) join these components by the shortest path between them; and (iii) update *dist* by running Dijkstra's algorithm once, with all vertices in the newly created component as sources (all other terminals must be reached). The heuristic runs in $O(|T|(|E| + |V| \log |V|))$ time, corresponding to $O(|T|)$ executions of Dijkstra's algorithm. We call this implementation Kruskal-B.

4.2   Improved Implementation (Kruskal-I)

We now propose an improved implementation, named Kruskal-I, that is much faster in practice. In each of its $|T| - 1$ iterations, Kruskal-I uses the Voronoi diagram associated with the partial solution S to identify the closest pair of components. We use the fact that there must be a frontier edge in the diagram that represents the shortest path in G between the closest pair of components. This is analogous to the the property stating that the Delaunay triangulation of a set of points contains the line segment connecting the two closest points (see [9], for instance).

The set of candidate shortest paths is in fact even more limited than that. Among all frontier edges (v, w) incident to v such that $v < w$, we define $\mathcal{E}(v)$ as the one whose associated path is the shortest (ties are broken arbitrarily). If there is no frontier edge (v, w) with $v < w$, we regard $\mathcal{E}(v)$ as undefined. Clearly, if the path associated with $\mathcal{E}(v)$ is not the shortest path in the graph, neither are the paths associated with the other frontier edges (v, w'), with $v < w'$.[3] This

---

[3] By considerig only edges in which v is the endpoint with the smallest label, we avoid associating the same edge to two different vertices.

definition ensures that there is always a vertex $v \in V$ such that $\mathcal{E}(v)$ represents the shortest path between the closest pair of components.

Kruskal-I maintains two data structures: the Voronoi diagram associated to the partial solution and a heap to select among candidate paths. It is actually a heap of *vertices*, since each vertex is mapped to a unique edge, which in turn is mapped to a path. The algorithm begins by calculating the Voronoi diagram associated with the initial solution, which contains all terminals and no edges. For each vertex v for which $\mathcal{E}(v)$ is defined, we insert v into the heap with cost equal to that of its associated path (if $\mathcal{E}(v) = (v, w)$, this cost is $d(v) + c(v, w) + d(w)$). The algorithm then starts adding paths to the solution. Each of its $|T| - 1$ iterations is divided into four steps:

1. *Determine the closest pair of components.* Remove the first element v from the heap, and let $\mathcal{E}(v) = (v, w)$ be its associated edge. If $base(v)$ and $base(w)$ both belong to the same component, then (v, w) is not actually a frontier edge (i.e., the value of $\mathcal{E}(v)$ is outdated).[4] In that case, search the neighborhood of v to determine the correct value of $\mathcal{E}(v)$ and reinsert v into the heap (if $\mathcal{E}(v)$ is not defined, disregard it). Proceed until a vertex associated with an actual frontier edge is removed. This vertex determines the closest pair of components (or, more accurately, it is associated with the frontier edge that represents the shortest path between the closest pair of components).

2. *Add to the partial solution the path joining the closest pair.* As in Borůvka's heuristic, use the Voronoi diagram to obtain the path from the frontier edge.

3. *Update the Voronoi diagram.* Run Dijkstra's algorithm having as sources all vertices newly inserted into the solution (also as in Borůvka's heuristic).

4. *Update the heap of candidates.* Each vertex v in the heap is associated with an edge $\mathcal{E}(v) = (v, v')$ and a value, the length of the path $\mathcal{E}(v)$ represents, given by $d(v) + c(v, v') + d(v')$. The information associated with v may become inaccurate (outdated) only if one or more of the following situations occur:
   (a) $d(v)$ is reduced;
   (b) $d(v'')$ is reduced, v'' being a neighbor of v (possibly v' itself);
   (c) (v, v') ceases to be a frontier edge, i.e., $base(v)$ and $base(v')$ become part of the same component in the solution.
   Note that the only vertices that may have their $d(\cdot)$ values reduced are those reached during the update of the Voronoi diagram (step 3). Therefore, it suffices to examine the neighborhood of those vertices to update the values associated to them (case (a)) and to their neighbors (case (b)). Although case (c) may involve other vertices, there is no need to address this case directly, since it can only cause an increase in the value associated to v. If we leave it unchanged, v will become the first element of the heap before it should. This is perfectly acceptable, as long as we are able to discover that (v, v') is not actually a frontier edge, and this is precisely what step 1 above does.[5] Note that cases (a) and (b) must be addressed as soon as they occur, since they potentially increase the priority of the vertices involved.

---

[4] Step 4 explains how the value may become outdated.

[5] There is also the possibility that v falls into cases (a) and (b) before it actually makes it to the top of the heap; it will be updated then.

Step 1 removes up to $O(|V|)$ elements from the heap, each in $O(\log|V|)$ time. Steps 1 and 4 may need to check the neighborhoods of all vertices, which requires $O(|E|)$ total time. Step 2 runs in time proportional to the number of vertices in the path $(O(|V|)$ in the worst case, considering the whole algorithm). Finally, updating the Voronoi diagram in step 3 requires $O(|E|+|V|\log|V|)$ time. Therefore, all $|T|-1$ iterations require $O(|T|(|E|+|V|\log|V|)$ time. This means Kruskal-I has the same worst-case complexity as Kruskal-B. However, Section 6 shows that Kruskal-I is much faster. This happens because the Voronoi diagram and the heap of candidates allow Kruskal-I to avoid useless, repeated operations. Updating these structures is on average much faster than the worst-case analysis suggest, since modifications tend to affect small portions of the graph.

5 Prim

Proposed by Takahashi and Matsuyama [11], the construtive heuristic for the SPG based on Prim's algorithm for the MST problem is probably the most commonly used in the literature [2,4,8,10]. The classic algorithm for the MST problem "grows" the solution from a root vertex. In each step, the vertex that is closer to the current partial solution is added to it, alongside with the connecting edge. After $|V|-1$ steps, all vertices are spanned. The corresponding heuristic for the SPG grows a tree starting from a vertex (usually a terminal). In each iteration, we add to the solution the closest terminal not yet inserted (alongside with the vertices in the path from the solution to that terminal). In $|T|-1$ iterations (or $|T|$, if the root is not a terminal), a valid Steiner tree is created.

5.1 Basic Implementation (Prim-B)

The basic implementation of the heuristic (which we call Prim-B) is discussed in [3,8], among others. It divides the algorithm into two phases. The first phase consists of building a $|V| \times |T|$ *table of distances*, which describes the shortest paths between each vertex and every terminal. In can be computed in $O(|T|(|E| + |V|\log|V|))$ time, by executing the Dijkstra's algorithm once from each terminal. We also maintain an auxiliary structure, called *closest*, which associates each terminal not yet inserted to the closest vertex in the partial solution. This structure can be initialized in $O(|T||V|)$ time: just scan the table of distances. The second phase is when the solution is actually built. In each of its $O(|T|)$ iterations, we scan *closest* to determine which terminal to add to the tree. The corresponding path (described in the table of distances) is then added to the solution. Finally, we update *closest*: for each vertex added to the solution, we check if its distance to any terminal not in S is less than what *closest* reports.

The bottleneck of the algorithm is building the table of distances. In [8], Polzin and Daneshmand suggest a Voronoi-based acceleration scheme that tries to expedite the process by limiting the extent of each execution of Dijkstra's algorithm. Our version of Prim-B tested in Section 6 includes such an acceleration. Although it does make the algorithm run faster for certain instances, it does not make it asymptotically more efficient on average.

5.2   Improved Implementation (Prim-I)

We now suggest an improved implementation (Prim-I) that works in a single phase. It can be seen as a modified version of Dijkstra's algorithm.

It starts with a partial solution S containing just the root r. To each vertex v_i we associate a variable π_i representing an upper bound to the distance from v_i to S in G. Initially, $\pi_r = 0$ (r is already in S) and $\pi_i = \infty$ for all vertices v_i not in S. In each iteration, the algorithm determines the closest terminal t and adds it to the partial solution, alongside with the non-terminals in the corresponding shortest path. The process of determining t is similar to a heap-based implementation of Dijkstra's algorithm.[6] Remove v_i, the element with the highest priority (smallest π_i), from the heap. If v_i is not a terminal or already belongs to the solution, then for each neighbor v_j of v_i, check whether $\pi_j < \pi_i + c(v_i, v_j)$. If that is the case, set $\pi_j \leftarrow \pi_i + c(v_i, v_j)$ and update the heap accordingly. Keep removing elements from the heap until a terminal not yet spanned is found. When this happens, add the corresponding new path to S.

The distance from any given vertex to the partial solution can only decrease in time. Hence, upper bounds found in any iteration remain valid until the end of the algorithm. To guarantee that the best vertices will be selected, all we have to do is reinsert every vertex v_i into the heap (with $\pi_i = 0$) as soon as it becomes part of the solution. This ensures that, in subsequent iterations, vertices that get closer to the solution will be visited if necessary.

The algorithm consists of $O(|T|)$ iterations ($|T|$ if the root is non-terminal, $|T| - 1$ if it is not). In each of these iterations, as many of $O(|V|)$ vertices may be removed from the heap, which takes $O(|V| \log |V|)$ time in the worst case. Each of them may have its neighborhood searched, and the values associated with the neighbors may need to be updated; this can be done in $O(|E|)$ time per iteration. Hence, the overall worst-case running time of Prim-I is $O(|T|(|E| + |V| \log |V|))$.

Note that some vertices may be inserted several times into the heap. However, this happens only if this is really necessary, that is, if their distances to the partial solution actually decrease. In fact, the experimental results in Section 6 suggest that, on average, a vertex is visited very few times during the execution of the algorithm. To understand why this happens, first note that each vertex is visited at most once per iteration, which means that no vertex can be visited more than $O(|T|)$ times. Furthermore, as the number of terminals increases, the number of vertices visited in each iteration tends to decrease. After all, with more terminals, there are on average fewer edges between the partial solution and the closest (non-inserted) terminal. It turns out that in practice these two factors (number of iterations and time for each iteration) balance each other, thus making the running time virtually independent of $|T|$.

This explains why Prim-I is much faster than Prim-B on average, although both have the same worst-case complexity. In fact, relying on a table of distances severely limits the applicability of the basic implementation. Merely initializing the table may be (and often is) much more expensive than running Prim-I in its entirety. But the main weakness of Prim-B is its quadratic memory usage; in practice, it cannot handle instances that are easily processed by Prim-I.

---

[6] It could be implemented without a heap, but it would not be as efficient in practice.

6   Empirical Analysis

This section presents an empirical comparison of the implementations discussed in this paper, with an emphasis on running times. Although we do compare the heuristics in terms of solution quality, we do it superficially. The reader is referred to [12] for a more thorough analysis, including not only most of the constructive heuristics discussed here (in their basic implementations), but also several other methods not directly based on algorithms for MST problems.

Test Problems. Our goal is to assess how the algorithms behave "on average". However, it is not clear how to do this strictly, since there are too many variables involved in a single instance of the problem (number of vertices, number of edges, graph topology, number of terminals, terminal distribution, and edge weights, to mention a few). We decided to use instances already available on the literature, assuming they represent a fair sample of what could be considered "typical" instances. The algorithms were tested on all 994 valid instances[7] available at the SteinLib repository [5]. To simplify the analysis, we grouped the original 36 series into six classes, as shown in Table 1.[8] These instances vary greatly in size. On average, they have 1,252 vertices, 9,800 edges, and 186 terminals; these dimensions reach values up to 38,418, 221,445, and 11,849, respectively.

Table 1. Instances available at the SteinLib

class	series	description
random	b, c, d, e, mc, p6z	random graphs, random weights
fst	es*fst, tspfst	rectilinear graphs, L1 weights
vlsi	alue, alut, dmxa, diw, gap, lin, msm, taq	grid graphs with holes
incidence	i080, i160, i320, i640	random graphs, incidence weights
euclidean	x, p6e	graphs with euclidean weights
hard	sp, puc	artificial, "hard" instances

Methodology. All algorithms were implemented in C++ and compiled under Linux with gcc 2.96 with the -O4 flag (full optimization). CPU times were obtained on a 1.7 GHz Pentium 4 with 256 MB of RAM. Because the timing function (getrusage) has low precision (1/60th of a second), running times could not be measured directly. Instead, each algorithm was repeatedly run on each instance until five seconds have passed, and the average time was considered.

All algorithms tested share the same implementation of basic data structures (graphs, heaps, stacks). We opted for using binary heaps to implement priority queues. Although they are asymptotically less efficient, their performance is competitive in practice with that of Fibonacci heaps [7]. With binary

---

[7] At the time of writing, September 1, 2001.

[8] In the table, the notion of "incidence weights" by that edges incident to terminals have larger weights than those incident to non-terminals (see [4]).

heaps, the worst cases of our algorithms are slightly different from those mentioned in previous sections. For DNH-Prim, DNH-Borůvka, and DNHz, the worst case is $O(|E| \log |V|)$; for Borůvka, $O(|E| \log |T| \log |V|)$; and for all versions of Prim and Kruskal, $O(|T||E| \log |V|)$. Note that these algorithms differ only in the dependency on $|T|$.

Although described as part of DNH, MST-prune may improve any valid solution, as noted in [12]. Therefore, unless otherwise noted, running times for all heuristics include this post-optimization phase, and solution qualities refer to results obtained *after* its application. MST-prune was implemented with Prim's algorithm (with binary heap), with roots selected at random. Its $O(|E| \log |V|)$ running time does not affect the worst-case complexities of the heuristics.

Comparing the performance of the heuristics with a well-known algorithm may convey a better assessment of the behavior of the methods discussed here. Therefore, our analysis includes an "extra" method, MST: it simply applies Prim's algorithm to calculate the minimum spanning tree of the entire graph (treating terminals as ordinary vertices). Running times for this method do not include MST-prune.

Results. Table 2 shows the average running times of each algorithm for all classes of instances. Due to its quadratic memory usage, implementation Prim-B could not be tested on 11 instances.[9] Some of the values in the tables below (marked with a "$\sim$" symbol) are therefore approximate, since they consider only the remaining instances.

Table 2. Average running times in milliseconds (with MST-prune)

method	complexity	euclidean	fst	hard	incidence	random	vlsi
DNH-Prim	$O(|E| \log |V|)$	2.71	1.95	3.68	4.51	3.41	11.02
DNH-Borůvka	$O(|E| \log |V|)$	3.27	1.99	**2.80**	3.42	2.80	**8.67**
DNHz	$O(|E| \log |V|)$	3.30	1.98	2.84	**3.41**	2.75	**8.67**
Borůvka	$O(|E| \log |V| \log |T|)$	3.36	2.19	3.21	3.57	3.00	12.65
Kruskal-B	$O(|T||E| \log |V|)$	49.25	270.70	589.78	399.53	162.48	237.54
Kruskal-I	$O(|T||E| \log |V|)$	5.22	3.07	4.26	3.70	4.11	16.72
Prim-B	$O(|T||E| \log |V|)$	106.73	$\sim$34.97	87.00	69.45	221.82	$\sim$36.73
Prim-I	$O(|T||E| \log |V|)$	**1.76**	**1.94**	3.97	6.67	**2.69**	8.86
MST	$O(|E| \log |V|)$	*1.36*	*0.69*	*1.74*	*2.90*	*1.84*	*5.89*

It should come as no surprise that DNH-Prim, DNH-Borůvka, and DNHz have average running times comparable to that of MST. They all have the same worst-case complexity, $O(|E| \log |V|)$, and share a very similiar structure. Due to the extra $\log |T|$ factor, $O(|E| \log |V|)$, Borůvka is a little slower, but still has comparable running times. The difference is much more evident for Kruskal-B and Prim-B, and can be explained by the $|T|$ factor in their complexity. However,

---

[9] Namely: alue7065, alue7080, alut2625, es10000fst01, fl3795fst, fnl4461fst, pcb3038fst, pla7397fst, rl11849fst, rl5915fst, and rl5934fst (first three belong to vlsi, others to fst).

the improved implementations of the same heuristics (Kruskal-I and Prim-I) also
have the extra $|T|$ factor, but are only slightly slower than DNH-Prim.

But being fast on average is not always enough. Several instances of the
SteinLib have hundreds or even thousands of terminals, and it could be the
case that the improved implementations perform poorly for these (and that the
average is small because of other instances, with few terminals). Table 3 shows
that this is not the case. For each instance, we calculated the *relative running
time* of each algorithm, that is, the ratio between its running time and that of
MST (the minimum spanning tree of the entire graph). The table shows the
best, average, and worst-case ratios obtained when all instances are considered.

Table 3. Relative running times for all instances (with MST-prune)

method	complexity	best	average	worst
DNH-Prim	$O(|E| \log |V|)$	1.072	2.076	4.364
DNH-Borůvka	$O(|E| \log |V|)$	0.645	1.762	5.840
DNHz	$O(|E| \log |V|)$	0.646	1.744	5.578
Borůvka	$O(|E| \log |V| \log |T|)$	0.668	1.995	5.992
Kruskal-B	$O(|T||E| \log |V|)$	3.135	56.184	1953.456
Kruskal-I	$O(|T||E| \log |V|)$	0.997	2.638	8.704
Prim-B	$O(|T||E| \log |V|)$	~2.597	~20.958	~838.426
Prim-I	$O(|T||E| \log |V|)$	0.778	2.154	6.567
MST	$O(|E| \log |V|)$	*1.000*	*1.000*	*1.000*

On average, none of the improved versions takes more than three times the
time required to calculate a simple minimum spanning tree. In some (rare) cases,
they can even be faster, because they do not necessarily have to examine the
whole graph. But, most importantly, Table 3 reveals that the worst cases of both
Prim-I and Kruskal-I are remarkably tolerable, especially when we consider that
more naive implementations of the same algorithms (Prim-B and Kruskal-B)
are up to hundreds of times slower than MST. Prim-I and Kruskal-I are never
more than ten times slower than MST. That is only twice as much as the fastest
heuristic tested, DNH-Prim, which has the same worst-case complexity as a
simple minimum spanning tree computation.

The results obtained so far strongly suggest that the performances of our
implementations do not depend as much on the value of $|T|$ as their worst-case
expressions indicate. In Figure 1, we show the relative running times of Prim-
I (without MST-prune, in this case) with respect to MST for all 994 Steinlib
instances. If both algorithms behaved in average as their worst-case complexities
suggest, the ratio should be linearly dependent on $|T|$. In log scale, this would
mean an exponentially-growing curve, but that is clearly not what we obtained.

Once again, this confirms what was said in Section 5: the worst-case analysis
is overly pessimistic when it assumes that each vertex is visited (i.e., removed
from the heap) once per iteration, or $\Theta(|T|)$ times during the whole algorithm.

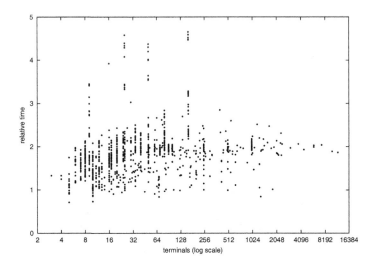

Fig. 1. Relative times of Prim-I for all SteinLib instances (without MST-prune)

For this particular set of instances, we verified that each vertex is visited only 2.17 times on average (recall that the average number of terminals is 186). Even in the worst case (instance i640-345, with 160 terminals), the average vertex was removed from the heap only 4.28 times.

Although all these experiments on SteinLib make it clear that Prim-I does not depend "too much" on $|T|$, it is not obvious what the dependence is. Because the set of instances is heterogeneous, there are simply too many variables (number of edges, number of vertices, graph topology, and so on) determining the results.

Therefore, we devised a more controlled experiment, where $|T|$ is the only relevant variable to be considered. While the results for this particular experiment cannot be directly extended to an any arbitrary set of instances, we believe it does help understand the behavior of the algorithms. We generated a new series, special, which may be seen as an extension of random. This is a series of random graphs, with integer edge weights uniformly distributed in $[1; 10]$ and random terminal placement. All graphs have exactly 1024 vertices and 32768 edges, and number of terminals ranging from 2 to 1024. For each value of $|T|$, there are 25 different graphs, generated with different random seeds. Figure 2 shows how the average running times of the algorithms depend on the size of T. In this particular test, all algorithms (including DNH) were run without MST-prune, thus allowing for a better assessment of the relative performance of each algorithm's "core".[10]

The figure shows that, for graphs in this series, only two methods (the basic implementations of Kruskal's and Prim's heuristics) actually depend linearly on

---

[10] For any given $|T|$, the effect of including MST-prune would be a uniform increase in running times for all heuristics. This effect is more noticeable for large values of $|T|$, but the additional running times are never greater than that of running MST.

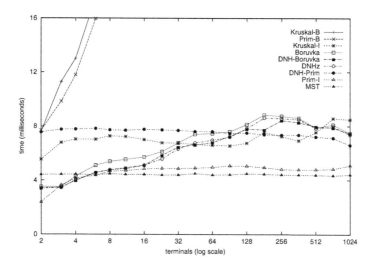

Fig. 2. Running times for series special (without MST-prune)

$|T|$, as the fast-growing curves indicate (note that the graph is in log scale). In fact, Kruskal-B takes 2.76 seconds on average for $|T| = 1024$ and Prim-B takes 0.78 seconds for $|T| = 724$ (Prim-B is actually faster than that for $|T| = 1024$ — just 0.09 seconds — because of the acceleration suggested in [8]). Both times are much higher than those of all other algorithms, including the new implementations of Kruskal and Prim. No other method takes more than 0.01 second to find a solution, regardless of the number of terminals.

Solution quality. Even with their best implementations, Prim and Kruskal are still a little slower on average than DNH, and, although we have established that the extra $|T|$ factor in their complexities does not tend to be relevant in practice, it does exist. All other things being equal, it seems that it would be faster and safer to use DNH rather than other heuristics. But there is something that is *not* equal: solutions provided by Prim's or Kruskal's heuristics are much better than those provided by DNH. Table 4 shows, for each class of instances, the average percentual error obtained by each algorithm (only one implementation of each algorithm is shown; others behave similarly to their counterparts). Errors were calculated with respect to the best solutions known on September 1, 2001 (available at the SteinLib).

The performance of each algorithm varies significantly among series. Within each series, however, the table clearly shows that the more actively an algorithm looks for "shortcuts" between terminals, the better it tends to behave with respect to other methods. DNH calculates the Voronoi diagram and determines all connections at once; it does not actively use previously inserted non-terminals to try to reduce the cost of the final solution. DNHz does that in a limited fashion, by favoring edges that are incident to vertices in the par-

Table 4. Average percentual error w.r.t. best known solution (with MST-prune)

method	euclidean	fst	hard	incidence	random	vlsi
DNH-Borůvka	2.53	2.70	24.23	23.31	5.03	5.24
DNHz	2.34	2.23	22.20	21.49	4.09	4.95
Borůvka	2.35	1.90	21.18	20.29	3.56	3.54
Kruskal-I	1.65	1.86	9.55	18.36	**2.68**	**2.56**
Prim-I	**1.28**	**1.83**	**7.87**	**17.92**	2.82	2.83

tial solution. Borůvka takes this strategy one step further: after each iteration, the whole Voronoi diagram is updated, so future links can take further advantange of the new non-terminal vertices in the solution. Kruskal also updates the Voronoi diagram after each iteration, but recall that an iteration in this algorithm corresponds to the addition of a single path, as opposed to multiple paths in Borůvka. This means that each decision made by Kruskal (and Prim) can take into account completely updated information about the partial solution. In general, this leads to solutions that are closer to the optimum.

7   Final Remarks

We have proposed new implementations of two important constructive heuristics for the Steiner Problem in Graphs: those based on Prim's and Kruskal's algorithms for the MST problem. Empirical results attest that these implementations are significantly faster and more robust than the usual, basic ones. We have also suggested some minor variations of Melhorn's implementation of DNH heuristic, shown to be competitive in practice. A Borůvka-based heuristic, which can be seen as an intermediate method between DNH and Kruskal (in terms of both solution quality and worst-case complexity), was also proposed.

From a theoretical point of view, an interesting extension of this work would be a strict average case analysis of the running times of the algorithms. However, this seems to be a non-trivial task, since there are too many variables to consider: $|V|$, $|E|$, $|T|$, edge weights, graph topology, terminal distribution, etc. Alternatively, one could identify specific classes of instances for which the new implementations (especially Kruskal-I and Prim-I) are guaranteed to perform fewer operations than their worst-case expressions suggest.

In terms of applications, it would be interesting to know to what extent these new implementations of constructive heuristics can help accelerate more elaborate exact and approximate algorithms for the SPG, as those described in [2,4,8]. In [10], the fast implementation of Prim's has already been used and greatly contributed to keep running times within acceptable limits.

Acknowledgements. We would like to thank Eduardo Uchoa, Diego Nehab, and two anonymous referees for their helpful comments. MPA was partially supported by CNPq, process 300475/93-4. RFW was partially supported by CAPES.

References

1. R. Ahuja, T. Magnanti, and J. Orlin. *Network Flows: Theory, algorithms, and applications.* Prentice-Hall, 1993.
2. C. Duin and S. Voss. The Pilot method: A strategy for heuristic repetition with application to the Steiner problem in graphs. *Networks*, 34:181–191, 1999.
3. F. Hwang, D. Richards, and P. Winter. *The Steiner tree problem*, volume 53 of *Annals of Discrete Mathematics*. North-Holland, Amsterdam, 1992.
4. T. Koch and A. Martin. Solving Steiner tree problems in graphs to optimality. *Networks*, 32:207–232, 1998.
5. T. Koch, A. Martin, and S. Voss. SteinLib: An updated library on Steiner tree problems in graphs. Technical Report ZIB-Report 00-37, Konrad-Zuse-Zentrum für Informationstechnik Berlin, 2000. `http://elib.zib.de/steinlib`.
6. K. Melhorn. A faster approximation algorithm for the Steiner problem in graphs. *Information Processing Letters*, 27:125–128, 1988.
7. B. M. E. Moret and H. D. Shapiro. An empirical assessment of algorithms for constructing a minimum spanning tree. *DIMACS Monographs in Discrete Mathematics and Theoretical Computer Science*, 15:99–117, 1994.
8. T. Polzin and S. V. Daneshmand. Improved algorithms for the Steiner problem in networks. *Discrete Applied Mathematics*, 112(1–3):263–300, 2001.
9. F. P. Preparata and M. I. Shamos. *Computational Geometry: An Introduction.* Springer-Verlag, 1985.
10. C. C. Ribeiro, E. Uchoa, and R. F. Werneck. A hybrid GRASP with perturbations for the Steiner problem in graphs. *INFORMS Journal on Computing*, to appear.
11. H. Takahashi and A. Matsuyama. An approximate solution for the Steiner problem in graphs. *Math. Japonica*, 24:573–577, 1980.
12. S. Voss. Steiner's problem in graphs: Heuristic methods. *Discrete Applied Mathematics*, 40:45–72, 1992.
13. S.M. Wang. A multiple source algorithm for suboptimum Steiner trees in graphs. In H. Noltemeier, editor, *Proceedings of the International Workshop on Graph-Theoretic Concepts in Computer Science*, pages 387–396. Würzburg, 1985.

A Time-Sensitive System for Black-Box Combinatorial Optimization

Vinhthuy Phan, Pavel Sumazin, and Steven Skiena*

State University of New York at Stony Brook, Stony Brook, NY 11794-4400 USA
{phan,psumazin,skiena}@cs.sunysb.edu

1 Introduction

When faced with a combinatorial optimization problem, practitioners often turn to black-box search heuristics such as simulated annealing and genetic algorithms. In *black-box optimization*, the problem-specific components are limited to functions that (1) generate candidate solutions, and (2) evaluate the quality of a given solution. A primary reason for the popularity of black-box optimization is its ease of implementation. The basic simulated annealing search algorithm can be implemented in roughly 30-50 lines of any modern programming language, not counting the problem-specific local-move and cost-evaluation functions. This search algorithm is so simple that it is often rewritten from scratch for each new application rather than being reused.

In this paper, we examine whether it pays to develop a more sophisticated, general-purpose heuristic optimization engine. The issue is whether a substantial performance improvement or ease-of-use gain results from using such an engine over the naïve implementation of search heuristics.

The case for well-engineered optimization engines has clearly been made for linear programming, where commercial LP packages (such as CPLEX) significantly outperform homegrown implementations of the simplex algorithm. The answer is not so clear for local search heuristics. A review [13] of six existing black-box optimization packages describes generally disappointing results. Indeed, one line of theoretical research has yielded the "no free lunch theorem" [19], suggesting that under certain assumptions the choice of search engine has no effect on the expected quality of the solution. Despite or perhaps because of this, an enormous variety of different heuristic search algorithms and variants have been proposed; see [1] for an excellent overview of this literature.

We believe that there is a role for a well-engineered heuristic search system, and this paper describes our preliminary efforts to build such a system. Table 1 provides at least anecdotal evidence of our success, by comparing the performance of our current combined search heuristic after one year of development to the tuned simulated annealing engine we developed after six months of work. The new engine produces substantially better results for all problems across almost all time scales.

---

* Supported in part by NSF Grant CCR-9988112 and ONR Award N00149710589.

D. Mount and C. Stein (Eds.): ALENEX 2002, LNCS 2409, pp. 16–28, 2002.
© Springer-Verlag Berlin Heidelberg 2002

Our paper is organized as follows. In Sect. 2, we outline some of the potential benefits which can accrue from building a well-engineered heuristic search system. In Sect. 3, we describe the general architecture of our prototype implementation *discropt*, which currently supports five different search heuristics to evaluate on each of six different combinatorial optimization problems. We then describe the search heuristics currently implemented in Sect. 4. In Sect. 5, we report on a series of experiments designed to compare the search heuristics on a wide variety of problems and time scales. We conclude with an analysis of these results and directions for further research.

Table 1. A comparison of the solution scores produced by our current combined heuristic with a previous version after six months of development. The better score appears in boldface.

Problem	Heuristic	1	5	10	15	20	25	30	45	60	90	120
VertCover	Early Version	2299	2260	2315	642	654	644	650	641	643	598	586
	Combination	**648**	**1100**	**571**	**554**	**541**	**546**	**528**	**531**	**521**	**521**	**523**
Bandwidth	Early Version	1970	**604**	**616**	**611**	**603**	**595**	622	620	609	**595**	**602**
	Combination	**693**	633	620	614	615	613	**609**	**607**	**608**	603	**602**
MaxCut	Early Version	**3972**	3972	3972	3972	3972	3969	3972	3972	3885	1373	1349
	Combination	**3972**	**1401**	**1385**	**1346**	**1462**	**1452**	**1441**	**1971**	**1689**	**1276**	**1240**
SCS	Early Version	3630	3688	1152	1158	1225	1120	1035	1013	994	902	982
	Combination	**1538**	**614**	**596**	**570**	**562**	**544**	**541**	**515**	**505**	**502**	**505**
TSP	Early Version	5101	1043	846	730	642	687	568	551	508	577	504
	Combination	**1133**	**933**	**633**	**472**	**411**	**406**	**397**	**378**	**381**	**375**	**374**
MaxSat	Early Version	27725	**572**	705	735	533	403	518	344	518	352	318
	Combination	**912**	636	**359**	**465**	**503**	**398**	**374**	**262**	**206**	**255**	**236**

2 Rationale

Many of the potential benefits of a general search system result from amortizing the cost of building a substantial infrastructure over many problems. These include:

- *Awareness of Time* – Running time is an important consideration in choosing the right heuristic for a given optimization problem. The choice of the right heuristic (from simple greedy to exhaustive search) depends upon how much time you are willing to give to it. A general search engine allows the user to specify the time allotted to the computation, and manages the clock so as to return the best possible answer on schedule.
  A time budget, constructed using an experimental determination of solution-evaluation and construction times, is required for proper time management. Once the budget and basic search landscape have been determined, a reasonable search strategy can be chosen. By monitoring progress and time as the

search unfolds we can judge the soundness of our selection and, if necessary, change strategies.

- *Ease of Implementation* – By incorporating objects for common solution representations (such as permutations and subsets) and interfaces for common input data formats, a generic system can significantly simplify the task of creating a code for a new optimization problem. Indeed, the six problems currently implemented in our system each require less than 50 lines of problem-specific code.
- *Inclusion of Multiple Heuristics* – A well-engineered heuristic search system can contain a wide variety of greedy, dynamic programming, and non-local search heuristics; uncovering solutions which would not be revealed just by local search. The cost of creating such an infrastructure cannot be justified for any single application. Our experience with generic greedy heuristics, reported in this paper, shows that they are valuable for a general search engine.
- *Parameter Tuning* – Search variants such as simulated annealing require a distressingly time-consuming effort to tune parameters such as the cooling schedule and distribution of moves. A well-engineered system can perform statistical analysis of the observed evaluation function values and automatically tune itself to obtain reasonable performance.
- *Testing and Evaluation Environment* – The choice of the best optimization engine for a given job is a difficult and subtle problem. Randomized search heuristics have a high variance in the quality of solutions found. Creating a proper testing and evaluation environment is a substantial enough task that it must be amortized over many problems. The reported success of many published heuristics is due to inadequate evaluation environments.

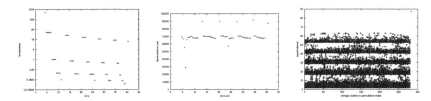

Fig. 1. Animating the fractional-restart search algorithm, including the temperature schedule (left), the evaluation rate (center), and positional hot spots vs. time (right).

- *Visualization Environment* – Animation tools which plot search progress as a function of time are very helpful in analyzing the behavior of the system and tuning parameters. Figures 1 and 4 show examples of plots built by our current implementation constructs, including (a) the value of the current solution as a function of time, (b) positions in the solution where changes proved beneficial, i.e. hot-spots, and (c) the number of evaluation function calls as a function of time, to measure the effect of system load over a long computation. This level of visualization support is difficult to justify building for each individual application.

- *Solution Quality Estimation* – A statistical analysis of the distribution of evaluated solution candidates can be used to estimate the number of solutions better than any given solution. Provided that the scores are normally distributed, this is a measure for the quality of the given solution. There is theoretical and experimental evidence to support the assumption that the distribution of scores is normal for many problems.

2.1 Literature

The literature on heuristic search is vast, encompassing dozens of books and a great number of papers. Heuristics such as *simulated annealing* [10], *genetic algorithms* [5] and *tabu search* [4] are particularly popular. We refer the interested reader to reviews of search heuristics for combinatorial optimization problems, including [1,2,20]. Our view on search algorithms has been shaped by our own combinatorial dominance theory of search, developed in [14].

Johnson, Aragon, McGeoch, and Schevon [8,9] conducted experimental evaluations of simulated annealing on graph partitioning, graph coloring and number partitioning. They found that simulated annealing was more effective on graph partitioning and coloring than number partitioning. They concluded that no variant of simulated annealing under consideration outperforms the rest across the board in any optimization problem. Our system provides a sound framework to conduct such systematic experiments.

A review of six black-box optimization systems is given by Mongeau, Karsenty, Rouzé, and Hiriart-Urruty [13]. They tested the engines on three types of problems: least median square regression, protein folding and multi-dimensional scaling, before concluding that no good self-adjusting black-box optimization engine is available in the public domain today. Most of the reviewed optimization engines [3,6,7,11,12,21] require a user adjustment of parameters, which heavily influences the engine's performance.

3   System Issues

In this section, we discuss the architecture of our prototype system and our test system, including a description of the optimization problems we evaluate it upon.

3.1   System Architecture

The general blueprint of our prototype system, written in C++, is given in the UML diagram in Figure 2. Our design consists of a central control unit, *solver*, that assigns work to the appropriate heuristic(s), and controls the information and statistics-gathering processes. The solver controls the:

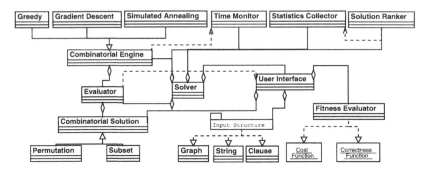

Fig. 2. The general blueprint of the system.

- *Combinatorial engine*, which is a generalization of our search heuristics; it forces a common interface on the heuristics and provides access mechanisms.
- *Evaluator*, which maintains a combinatorial solution, and information related to the solution (most notably its cost, correctness and state).
- *Combinatorial solution*, which is a generalization of permutation and subset solution types.
- *Monitor*, which maintains real time and system time information; it notifies the engine when to report solutions, when to evaluate its position in the search process and when to terminate.
- *Statistics collector*, which gathers information about cost function evaluation time (including partial solution evaluation), search progress in the form of solution evaluation snapshots, score distribution and more.
- *User interface*, which is the only component accessible to system users; it is used to define cost functions and specify the solution type.

User-defined cost functions include a complete solution cost function (see example in Figure 3), and (optionally) correctness, partial cost, and partial correctness functions. The partial-cost function is particularly useful for a fast evaluation of solutions related to previously-evaluated solutions. An example is the partial-cost function for $Swap$: given a previously evaluated solution s_0, its swap-neighbor s_1, and the cost of $Swap(s_0, s_1)$, the user can define the operation $\oplus$ so that $\text{cost}(s_1) = \text{cost}(s_0) \oplus \text{cost}(Swap(s_0, s_1))$. In the case of TSP, the partial solution is evaluated in constant time, while a complete solution is evaluated in $O(n)$ time.

3.2   Test Problems

We have carefully selected a suite of combinatorial optimization problems for their diverse properties. Our primary interest is in the performance of the search heuristics across a general spectrum of problems. Due to lack of space, we limit the discussion to one instance for each problem. However, we assure the reader that the reported results are representative of the other instances in our test bed. Indeed, our full experimental results are available at [15].

```
double UserInterface::cost_function(CombinatorialSolution & solution)
{
 double weight=0;
 solution_index first=solution.get_first_element(), last=solution.get_last_element(), i;
 element next, current = solution.element_lookup(first);
 for(i=first; i<last; current=next, i++){
   next = solution.element_lookup(solution.next_element(i));
   weight += input->edge_weight(current, next);
 }
 weight += input->edge_weight(solution.element_lookup(last), solution.element_lookup(first));
 return weight;
}
```

Fig. 3. An example of cost function for the Traveling Salesman Problem.

- *Bandwidth* - given a graph, order the vertices to minimize the maximum difference between the positions of the two vertices defining each edge. Bandwidth is a difficult problem for local search, because the solution cost depends completely on the maximum stretch edge. The representative test case is file alb2000 from TSPlib [16,17], a graph of 2000 vertices and 3996 edges. The number of C++ lines of code used to define the complete and partial cost functions are, respectively, 9 and 30.
- *Traveling Salesman* - given an edge-weighted graph, find a shortest tour visiting each vertex exactly once. TSP is well suited to local search heuristics, as move generation and incremental cost evaluation are inexpensive. The representative test case is a complete graph on 666 vertices from TSPlib [16], gr666. Lines of code: 9 and 23.
- *Shortest Common Superstring* - given a set of strings, find a shortest string which contains as substrings each input string. SCS can be viewed as a special case of directed, non-Euclidean TSP, however here it is deliberately implemented by explicitly testing the maximum overlap of pairs of strings on demand. This makes the evaluation function extremely slow, limiting the performance of any search algorithm. Our test case consists of 50 strings, each of length 100, randomly chosen from 10 concatenated copies of the length-500 prefix of the Book of Genesis. Lines of code: 13 and 23.
- *Max Cut* - partition a graph into two sets of vertices which maximize the number of edges between them. Max cut is a subset problem which is well suited to heuristic search. The representative test case is the graph alb4000 from TSPlib [16] with 4000 vertices and 7997 edges. Lines of code: 10 and 17.
- *Max Satisfiability* - given a set of weighted Boolean clauses, find an assignment that maximizes the weights of satisfiable clauses. Max sat is a difficult subset problem not particularly well suited to local search. The representative test case is the file jnh10 from [18], a set of 850 randomly weighted clauses of 100 variables. Lines of codes: 12 and 14.
- *Vertex Cover* - find a smallest set of vertices incident on all edges in a graph. Vertex cover is representative of constrained subset problems, where we seek the smallest subset satisfying a correctness constraint. The changing impor-

tance of size and correctness over the course of the search are a significant complications for any heuristic. The representative test case is a randomly generated graph of 1000 vertices and 1998 edges. Lines of code: 8 and 2 for cost, 19 and 2 for correctness.

4   Search Algorithms

We present two fundamentally different approaches for combinatorial optimization, specifically variants of simulated annealing – a neighborhood search technique, and incremental greedy heuristics. A neighborhood search heuristic traverses a solution space S, using a *cost function*, $f : S \to \Re$, moving from one solution to another solution in its *neighborhood* $N : S \to 2^S$. We use the *swap* operation as our neighborhood operator. In the case of permutation solutions, a swap exchanges the positions of two permutation elements; in the case of subset solutions, a swap operation moves a given element in or out of the subset. For certain problems, the solution space may contain invalid solutions. It is then necessary to have a *correctness function*, $c : S \to \Re$, which indicates a solution's degree of validity. A non-neighborhood search method such as *incremental greedy* constructs solutions by incrementally adding solution fragments.

4.1   Simulated Annealing

Simulated annealing is a gradient descent heuristic that allows backward moves at a rate that is probabilicaly proportional to $e^{-\frac{\Delta}{c}}$, where Δ is the difference in score between two solutions and c is the current temperature. The higher the temperature, the more likely the acceptance of a backward move. The initial temperature is set high to allow wide solution-space exploration, and is cooled down as the search progresses, eventually leading to a state where no backward moves are accepted. At each temperature, the search explores L_k states or solutions. Performance of a simulated annealing strategy depends its cooling strategy. Our system includes three variants of simulated annealing; typical search progress for these variants is given in Figure 4.

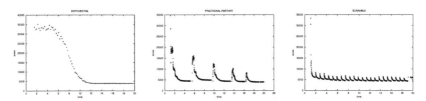

Fig. 4. Solution score vs. time for three different simulated annealing schedules: exponential (left), fractional-restart (center), and scramble (right).

One-run Simulated Annealing. This strategy reduces its temperature from the highest to the lowest once, using the full running time. After L solutions are explored, the current temperature c is replaced by $c \cdot \Delta c$, for some Δc computed by (1) estimating the number of decrements of temperature, $k = \frac{T}{\Delta t}$, where T is the remaining time; Δt is the time to explore L solutions, and (2) letting the search reach the estimated lowest temperature. Holding L and therefore Δt constant, we can deduce Δc. For greater time accuracy, Δc is updated dynamically each time the temperature is reduced.

When Δt is too large in comparison to T, k is too small to allow for a smooth temperature reduction. When Δt is too small in comparison to T, temperature may get decreased too quickly, leading to a premature termination. To balance these extremes, L assumes a small initial value and is increased gradually.

Multiple-restart Simulated Annealing. The multiple-restart strategy consists of two steps: first converge quickly to a local optimum; second make a number of backward moves in the vicinity of that local optimum. We study two restart variants. *Fractional restart* resets the initial temperature to a fraction of the previous initial temperature; the fraction is computed so that at least one restart trial is carried out before time runs out. *Scrambling* explicitly *backtracks* from the local optimum; the backtracking distance is a predetermined fraction of the previous restart.

4.2 General Greedy Heuristics

The greedy heuristic constructs a solution by incrementally adding elements to a partial solution. At each increment, greedy selects the best of k candidate elements for inclusion; k is determined by the specified running time. For ordered solutions such as permutations, the system includes two greedy variants: (1) *interleave* places the best candidate at the best solution position, and (2) *append* appends the best candidate to the existing solution. For subset problems, there is no distinction between the two heuristics.

4.3 A Combined Strategy

The performance of search strategies vary across different problems, and across different problem instances. The *combined* strategy allots each heuristic a fraction of the time, then chooses the best candidate heuristic based on the information gathered from the initial runs. While a heuristic's performance after a short running time need not be indicative of its performance over a longer time, in practice we are able to predict whether greedy will perform better than simulated annealing on a vast majority of our test instances.

4.4 Engine Efficiency

While some system and optimization overhead is unavoidable, an engine that spends only a small portion of its execution time on solution evaluation is doomed to poor performance. We evaluate the overhead of each of our heuristics in Table 2.

The proportion of time spent examining new solutions depends on the following factors:

- *The amount of overhead done by the engine* – this includes initialization, recalibration, statistics gathering, and time management. For example, if the specified running time is barely larger than the time needed for initialization, the proportion of time spent on evaluation must be small. The greedy heuristic spends more than 95% of the time in evaluation of shortest common superstring solutions over 5 seconds, but less than 50% when the running time is under 1 second.
- *The evaluation time for each solution* – much of the overhead work is not instance dependent, therefore when evaluation is costly, the proportion of the time spent on evaluation is larger. The evaluation function for shortest common superstring is more time-intensive than the evaluation of a comparable size traveling-salesman solution, and the proportion of time spent on evaluation is accordingly greater.
- *The frequencies of partial and complete solution evaluations* – the evaluation time for each solution is dependent on the evaluation method. Partial evaluation is often less time-intensive than complete evaluation. The interleave heuristic performs many partial solution evaluations, while the append heuristic performs mostly complete solution evaluations for permutation based solutions. Thus, for instances of bandwidth and TSP the portion of time spent on evaluation by append is much larger than that of interleave.

Table 2. The percentage of time spent on evaluating solutions. The instances are the same as those reported in Fig. 5. The *append* and *interleave* heuristics are identical for *subset* problems (maxcut, maxsat, and vertex cover) as discussed previously. Note that some solution evaluations are used in the initialization phase and do not contribute to search progression.

Program	Append	Interleave	Exponential	Frac. Restart	Scramble	Combination
Bandwidth	81.9%	42.1%	58.7%	55.0%	51.7%	63.0%
TSP	80.7%	55.1%	60.5%	62.0%	56.1%	58.4%
SCS	96.5%	94.8%	97.0%	97.4%	96.9%	94.9%
MaxCut	34.7%	34.7%	44.9%	46.2%	44.3%	42.3%
maxSat	97.8%	97.8%	97.2%	97.6%	97.3%	98.1%
VertCover	51.5%	51.5%	43.4%	51.1%	51.2%	55.0%

5   Experimental Results

Our experimental results are given in Fig. 5 and Table 3. The experiments were carried out on an AMD 1Ghz personal computer running Linux with 768MB of RAM. Each heuristic was performed 10 times for each heuristic on running times ranging from 1 to 120 seconds.

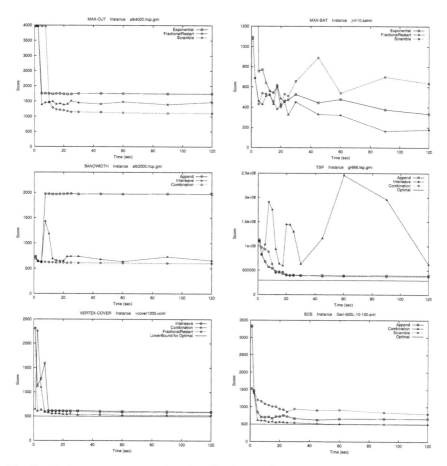

Fig. 5. Best score versus running time. In the top figures we compare the simulated annealing heuristics; in the middle figures we compare the greedy heuristics; and in the bottom figures we compare the best performing simulated annealing heuristics and the best performing greedy heuristics. For complete results see Table 3.

For each instance/heuristic, there is a time threshold below which its performance is uncharacteristically bad. When given a running time below this threshold, the heuristic aborts its intended strategy and resorts to simple gradient descent search. For example, note the performance of interleave on the bandwidth and TSP instances in Fig. 5. Some heuristics perform well for linear objective functions but not for non-linear objective function. For example, observe the behavior of append on bandwidth. Our experiments suggest that greedy

Table 3. Mean and standard deviation of scores at given running-time. Scores with the best mean, and those whose means are smaller than the sum of the best mean and its standard deviation appear in boldface. Results for TSP are given in units of 1000.

Problem	Time	Append Mean	Stdv	Interleave Mean	Stdv	Combination Mean	Stdv	Exponential Mean	Stdv	FractionalRestart Mean	Stdv	Scramble Mean	Stdv
SCS	1	3338.6	91.8	3327.1	128.5	1538.3	109.0	1521.6	70.7	3672.4	23.6	3314.0	159.6
	2.5	1476.7	95.5	1506.7	93.4	1398.4	67.7	1413.5	44.0	3617.3	105.6	1438.0	68.3
	5	840.5	101.7	1554.7	212.9	614.8	55.2	1500.8	115.3	3585.2	96.1	1198.3	119.3
	7.5	705.2	125.8	1202.7	87.9	598.1	38.9	1392.1	122.2	3609.7	69.4	1161.2	88.2
	10	699.9	125.0	1116.4	72.3	596.9	23.0	1339.9	35.8	1378.2	104.1	1087.8	92.7
	12.5	709.9	50.9	1052.0	129.5	562.2	54.1	1265.7	162.5	1132.4	97.0	1050.4	116.7
	15	668.7	85.1	917.6	84.9	570.7	45.3	1129.3	118.1	1135.7	121.3	1009.4	106.5
	20	718.4	94.8	996.0	117.8	562.2	42.2	1059.4	111.4	1037.9	95.3	940.2	94.2
	25	724.9	77.5	823.3	114.6	544.0	42.8	1085.4	126.1	1006.4	143.1	854.2	51.7
	30	663.9	90.9	889.6	100.1	541.7	38.1	1085.8	124.4	1024.4	83.5	938.3	71.1
	45	621.6	77.0	868.6	129.9	515.2	29.6	1048.7	105.9	875.7	115.7	895.5	78.4
	60	654.0	85.5	839.9	62.8	505.4	10.8	946.0	57.1	880.8	90.6	909.9	71.1
	90	660.1	55.2	847.8	101.6	502.7	8.1	902.1	103.9	823.5	127.8	839.3	148.2
	120	657.3	66.7	777.2	132.8	505.4	10.8	878.4	120.4	844.2	88.2	798.1	119.2
Bandwidth	1	731.3	11.4	739.2	16.0	693.5	7.8	660.4	12.8	1967.3	10.4	734.3	11.4
	2.5	643.2	6.6	643.5	13.5	644.2	9.1	632.2	11.6	1973.4	13.5	645.2	8.3
	5	624.8	9.9	622.7	9.5	633.1	6.8	615.7	8.6	790.8	17.7	619.9	11.3
	7.5	1974.6	12.8	1428.5	14.1	627.3	9.5	615.7	9.2	641.7	11.5	614.1	11.7
	10	1973.9	10.6	1181.0	18.5	620.9	9.6	611.5	8.4	616.5	8.9	615.6	14.3
	12.5	1968.6	13.9	691.7	17.3	619.8	8.8	734.1	381.0	606.1	10.3	613.7	8.3
	15	1969.3	11.7	659.6	16.9	614.8	8.2	606.8	9.5	606.3	8.6	613.4	13.7
	20	1972.1	11.6	646.1	9.7	615.5	7.2	982.5	559.7	602.6	9.3	610.8	13.7
	25	1979.0	6.9	737.2	16.2	613.0	10.9	609.4	14.9	603.2	10.9	604.7	10.6
	30	1969.7	14.6	737.0	16.0	609.9	5.0	483.5	3.7	608.4	7.6	606.0	9.4
	45	1973.6	10.8	678.5	13.5	607.9	4.8	482.2	5.6	605.2	7.3	608.4	8.9
	60	1971.9	14.6	635.9	19.7	608.2	3.7	477.9	4.5	599.6	9.7	605.7	9.0
	90	1976.0	10.5	729.9	12.7	602.4	7.1	462.5	5.4	600.9	9.0	603.6	8.0
	120	1972.4	12.6	656.3	10.9	602.2	5.8	453.0	3.4	596.8	13.9	606.2	16.4
MaxCut	1	3972.0	0.0	3972.0	0.0	3972.0	34.7	3972.0	0.0	3972.0	0.0	3972.0	0.0
	2.5	3972.0	0.0	3972.0	0.0	3971.7	31.5	3972.0	0.0	3972.0	0.0	3972.0	0.0
	5	3963.8	15.0	3963.8	15.0	1401.8	14.1	1754.4	25.0	1387.9	20.9	3969.9	6.0
	7.5	3964.4	12.0	3964.4	12.0	1354.0	26.2	1756.8	19.2	1453.6	52.9	3965.9	17.3
	10	1575.4	20.7	1575.4	20.7	1385.5	53.1	1751.9	27.5	1455.0	72.5	1467.6	120.0
	12.5	1321.1	36.9	1321.1	36.9	1355.2	26.8	1740.0	21.6	1492.5	222.6	1288.5	97.7
	15	1224.7	16.2	1224.7	16.2	1346.7	18.9	1747.2	29.4	1402.1	61.3	1227.5	15.3
	20	1221.4	16.6	1221.4	16.6	1462.9	64.5	1738.4	31.0	1385.2	69.1	1199.6	8.8
	25	1167.2	19.7	1167.2	19.7	1452.0	76.6	1733.0	22.6	1512.6	102.9	1146.9	6.2
	30	1150.3	12.8	1150.3	12.8	1441.6	46.0	1753.9	25.3	1450.2	56.3	1144.9	10.6
	45	1662.2	27.9	1662.2	27.9	1971.2	12.8	1755.3	23.5	1409.8	51.3	1140.6	13.5
	60	1630.0	42.5	1630.0	42.5	1689.1	90.2	1751.5	13.7	1480.5	262.1	1123.2	13.8
	90	1281.4	35.7	1281.4	35.7	1276.6	36.0	1746.0	33.5	1394.3	49.0	1109.5	11.9
	120	1278.2	51.9	1278.2	51.9	1240.0	22.6	1748.7	24.1	1468.7	219.1	1104.6	13.1
VertCover	1	2313.90	20.33	2313.90	20.33	648.49	5.56	648.90	4.21	2307.37	29.31	2324.30	11.93
	2.5	1123.29	766.20	1123.29	766.20	2259.57	23.77	650.00	7.58	606.80	6.94	616.90	4.87
	5	1272.97	817.17	1272.97	817.17	1100.23	760.28	650.70	4.80	633.17	3.71	604.80	4.66
	7.5	1592.26	808.24	1592.26	808.24	582.28	15.23	649.20	4.83	602.50	5.30	599.10	5.34
	10	617.17	5.91	617.17	5.91	571.32	34.84	653.00	8.41	603.20	4.69	632.30	11.40
	12.5	615.47	5.50	615.47	5.50	550.75	15.64	647.70	7.54	600.50	3.69	626.80	9.10
	15	613.43	3.43	613.43	3.43	554.95	17.59	646.80	6.00	601.00	3.03	621.40	5.06
	20	614.92	3.39	614.92	3.39	541.86	13.98	651.50	5.70	599.00	4.22	625.60	5.39
	25	614.95	5.38	614.95	5.38	546.13	13.80	650.60	3.85	597.50	3.64	621.10	6.73
	30	614.89	5.95	614.89	5.95	528.80	12.39	649.40	5.22	597.50	5.43	615.80	4.31
	45	604.80	4.02	604.80	4.02	531.68	13.42	604.50	5.95	584.40	3.04	609.60	2.33
	60	602.90	1.92	602.90	1.92	521.49	10.82	600.40	3.88	581.20	2.96	606.10	3.83
	90	596.90	4.04	596.90	4.04	521.89	9.54	594.90	4.89	579.30	1.79	598.40	3.69
	120	594.20	2.75	594.20	2.75	523.05	9.03	591.80	16.58	580.90	2.66	598.40	4.43
TSP	1	1102.0	42.8	1100.7	38.5	1133.4	39.1	1074.8	32.2	5085.0	30.6	1093.5	33.4
	2.5	825.8	31.9	825.4	34.6	971.0	36.2	908.0	29.5	1339.3	39.9	823.3	19.1
	5	680.6	42.8	1033.9	29.9	933.7	52.3	883.0	19.2	835.3	27.9	676.4	30.4
	7.5	570.5	27.8	1907.7	139.2	885.3	16.4	725.8	23.3	781.5	46.8	627.1	23.2
	10	539.7	32.0	1747.8	77.3	633.3	189.8	769.0	64.1	716.7	29.0	635.2	62.7
	12.5	476.3	12.7	936.3	60.5	448.1	13.6	697.3	29.6	698.3	22.8	667.8	67.3
	15	459.3	18.4	636.9	24.0	472.9	129.0	672.8	39.7	667.7	26.8	676.4	73.2
	20	404.4	16.1	1445.0	167.3	411.5	12.5	614.1	33.4	617.3	31.5	693.4	28.1
	25	397.8	4.9	1295.6	98.6	406.1	13.6	591.3	27.5	578.8	14.3	658.9	17.7
	30	390.8	10.3	633.2	36.4	397.4	18.0	573.4	11.2	544.0	16.6	659.8	27.7
	45	393.9	9.5	1159.4	169.0	378.7	5.6	524.6	17.5	509.3	14.2	630.4	23.5
	60	390.4	5.5	2457.0	661.2	381.0	6.0	488.2	14.2	529.0	30.8	615.5	19.6
	90	393.6	5.0	1955.5	276.0	375.8	5.0	472.2	14.4	487.1	21.8	591.2	22.0
	120	393.6	6.8	622.4	51.7	374.5	3.3	449.5	15.2	475.9	13.1	569.5	23.5
MaxSat	1	877.9	378.8	877.9	378.8	912.3	320.1	3119.2	747.6	1074.9	285.8	1092.3	473.3
	2.5	674.5	215.6	674.5	215.6	779.3	169.9	3668.5	488.8	685.9	265.8	686.2	285.2
	5	473.2	282.3	473.2	282.3	636.4	237.7	755.8	193.7	459.5	208.7	431.5	198.9
	7.5	530.4	179.0	530.4	179.0	601.9	270.3	773.2	318.7	428.6	199.1	539.2	230.7
	10	443.0	159.0	443.0	159.0	359.2	114.7	642.0	188.9	509.5	234.9	531.1	238.5
	12.5	510.9	184.8	510.9	184.8	493.5	212.2	559.4	179.4	526.2	144.9	520.1	183.9
	15	448.9	214.5	448.9	214.5	465.5	159.1	543.5	233.2	429.9	213.5	458.8	179.7
	20	402.4	198.6	402.4	198.6	503.4	211.6	401.3	127.9	469.3	354.2	426.8	169.3
	25	685.3	424.3	685.3	424.3	398.9	210.6	474.1	230.4	325.6	158.1	510.0	290.3
	30	644.9	230.1	644.9	230.1	374.1	142.8	526.1	169.3	451.5	147.8	656.3	364.8
	45	796.3	390.2	796.3	390.2	262.1	154.5	444.9	214.2	331.1	166.7	889.3	299.2
	60	719.0	385.2	719.0	385.2	206.5	101.9	477.7	157.9	321.4	102.7	538.6	265.7
	90	700.3	147.0	700.3	147.0	255.8	97.6	375.9	108.0	161.7	84.2	698.8	162.8
	120	560.0	284.0	560.0	284.0	236.2	78.7	335.1	82.0	179.4	123.3	635.8	321.3

heuristics are useful in the context of a time-sensitive combinatorial system, and that a combination of greedy and simulated annealing often outperforms either stand-alone component heuristic.

Results for one instance of each problem are presented, but the analyzed behavior appears representative of all problem instances (see [15]). The results support the following claims:

- No single heuristic outperforms the rest across all six problems.
- Given a collection of heuristics, there is an advantage in combining and selecting the best heuristics for a given problem or instance. Indeed, our combined heuristic performs significantly better than the five pure heuristics as reported in Table 3. The combined heuristic produces the best results for at least some time interval for every problem, and for every time interval on some problem.
- Greedy heuristics do better than simulated annealing when time is short, but for longer running times, the simulated annealing heuristics will excel even on problems that have constant-factor greedy approximation algorithms, such as vertex cover. For example, simulated annealing does not perform as well as greedy on our traveling salesman instance within the 120 seconds allotted. However, given 3 hours, the exponential simulated annealing heuristic finds a solution of score 326828, which is within near optimal optimal and far better than the solution found by the greedy heuristic.
- Greedy heuristics and other constructive strategies do better than local search when energy landscapes are relatively flat; for example, shortest common superstring.

References

[1] E. Aarts and J.-K. Lenstra. *Local Search in Combinatorial Optimization*. Wiley-Interscience, Chichester, England, 1997.

[2] D. Corne, M. Dorigo, and F. Glover. *New Ideas in Optimization*. McGraw-Hill, London, 1999.

[3] T. Csendes and D. Ratz. Subdivision direction selection in interval methods for global optimization. *SIAM Journal on Numerical Analysis*, 34(3):922–938, 1997.

[4] F. Glover. Tabu search– part I. *ORSA Journal on Computing*, 1(3):190–206, 1989.

[5] J. H. Holland. *Adaptation in natural artificial systems*. University of Michigan Press, Ann Arbor, 1975.

[6] L. Ingber. Adaptive simulated annealing (asa): Lessons learned. *Control and Cybernetics*, 25(1):33–54, 1996.

[7] M. Jelasity. Towards automatic domain knowledge extraction for evolutionary heuristics. In *Parallel Problem Solving from Nature - PPSN VI, 6th International Conference*, volume 1917 of *Lecture Notes in Computer Science*, pages 755–764, Paris, France, Sept. 2000. Springer.

[8] D. S. Johnson, C. R. Aragon, L. A. McGeoch, and C. Schevon. Optimization by simulated annealing: an experimental evaluation; part 1, graph partitioning. *Operations Research*, 37(6):865–892, 1989.

[9] D. S. Johnson, C. R. Aragon, L. A. McGeoch, and C. Schevon. Optimization by simulated annealing: an experimental evaluation; part 2, graph coloring and number partitioning. *Operations Research*, 39(3):878–406, 1991.

[10] S. Kirpatrick, C. Gelatt, Jr., and M. Vecchi. Optimization by simulated annealing. *Science*, 220:671–680, May 1983.

[11] A. V. Kuntsevich. Fortran-77 and fortran-90 global optimization toolbox: User's guide. Technical Report A-8010, Institut fur Mathematic, Karl Franzens Universitat, Graz, Austria, 1995.

[12] L. Lukšan, M. Tůma, M. Šiška, J. Vlček, and N. Ramešová. Interactive system for universal functional optimization (ufo). Technical Report 826, Institute of computer science, Academy of sciences of the Czech Republic, Prague, Czech Republic, 2000.

[13] M. Mongeau, H. Karsenty, V. Rouzé, and J.-B. Hiriart-Urruty. Comparison of public-domain software for black box global optimization. *Optimization Methods and Software*, 13(3):203–226, 2000.

[14] V. Phan, S. Skiena, and P. Sumazin. A model for analyzing black box optimization. in preparation, 2001.

[15] V. Phan, P. Sumazin, and S. Skiena. Discropt web page. http://www.cs.sunysb.edu/~discropt.

[16] G. Reinelt. *TSPLIB*. University of Heidelberg, www.iwr.uni-heidelberg.de/groups/comopt/software/TSPLIB95.

[17] G. Reinelt. TSPLIB— A traveling salesman problem library. *ORSA Journal on Computing*, 3(4):376–384, 1991.

[18] M. Resende. *Max-Satisfiability Data*. Information Sciences Research Center, AT&T, www.research.att.com/~mgcr.

[19] D. H. Wolpert and W. G. Macready. No free lunch theorems for optimization. *IEEE Transactions on Evolutionary Computation*, 1(1):67–82, 1997.

[20] M. Yagiura and T. Ibaraki. On metaheuristic algorithms for combinatorial optimization problems. *The Transactions of the Institute of Electronics, Information and Communication Engineers*, J83-D-1(1):3–25, 200.

[21] Q. Zheng and D. Zhuang. Integral global optimization: Algorithms, implementations and numerical tests. *Journal of Global Optimization*, 7(4):421–454, 1995.

A Compressed Breadth-First Search for Satisfiability

DoRon B. Motter and Igor L. Markov

Department of EECS, University of Michigan, 1301 Beal Ave, Ann Arbor, MI 48109-2122
{dmotter, imarkov}@eecs.umich.edu

Abstract. Leading algorithms for Boolean satisfiability (SAT) are based on either a depth-first tree traversal of the search space (the DLL procedure [6]) or resolution (the DP procedure [7]). In this work we introduce a variant of Breadth-First Search (BFS) based on the ability of Zero-Suppressed Binary Decision Diagrams (ZDDs) to compactly represent sparse or structured collections of subsets. While a BFS may require an exponential amount of memory, our new algorithm performs BFS directly with an implicit representation and achieves unconventional reductions in the search space.
We empirically evaluate our implementation on classical SAT instances difficult for DLL/DP solvers. Our main result is the empirical $\Theta(n^4)$ runtime for hole-n instances, on which DLL solvers require exponential time.

1   Introduction

Efficient methods to solve SAT instances have widespread applications in practice and have been the focus of much recent research [16,17]. Even with the many advances, a number of practical and constructed instances remain difficult to solve. This is primarily due to the size of solution spaces to be searched.

Independently from SAT, [9] explored Lempel-Ziv compression in exhaustive search applications such as game-playing and achieved memory reductions by a small constant factor. Text searches in properly indexed databases compressed using a Burrows-Wheeler scheme were proposed in [8]. However, these works enable only a limited set of operations on compressed data, and the asymptotic compression ratios are too small to change the difficulty of search for satisfiability. A different type of compression is demonstrated by Reduced Ordered Binary Decision Diagrams (ROBDDs) and Zero-Suppressed Binary Decision Diagrams (ZDDs) [13]. In general, Binary Decision Diagrams implicitly represent combinatorial objects by the set of paths in a directed acyclic graph. Complexity of algorithms that operate on BDDs is often polynomial in the size of the BDD. Therefore, by reducing the size of BDDs, one increases the efficiency of algorithms. There have been several efforts to leverage the power of BDDs and ZDDs as a mechanism for improving the efficiency of SAT solvers. [4] [5] used ZDDs to store a clause database and perform DP on implicit representations. [2] used ZDDs to compress the containers used in DLL. Another method is to iteratively construct the BDD corresponding to a given CNF formula [19]. In general this method has a different class of tractable instances than DP [11]. Creating this BDD is known to require exponential time for several instances [11], [19]. For random 3-SAT instances, this method also gives slightly different behavior than DP/DLL solvers [12].

D. Mount and C. Stein (Eds.): ALENEX 2002, LNCS 2409, pp. 29–42, 2002.
© Springer-Verlag Berlin Heidelberg 2002

The goal of our work is to use ZDDs in an entirely new context, namely to make Breadth-First Search (BFS) practical. The primary disadvantage of BFS — exponential memory requirement — often arises as a consequence of explicit state representations (queues, priority queues). As demonstrated by our new algorithm Cassatt, integrating BFS with a compressed data structure significantly extends its power. On classical, hard SAT benchmarks, our implementation achieves asymptotic speed-ups over published DLL solvers. To the best of our knowledge, *this is the first work in published literature to propose a compressed BFS as a method for solving the SAT decision problem*. While Cassatt does not return a satisfying solution if it exists, any such SAT oracle can be used to find satisfying solutions with at most $|Vars|$ calls to this oracle.

The remaining part of the paper is organized as follows. Section 2 covers the background necessary to describe the Cassatt algorithm. A motivating example for our work is shown in Section 3. In Section 4 we discuss the implicit representation used in Cassatt, and the algorithm itself is described in Section 5. Section 6 presents our experimental results. Conclusions and directions of our ongoing work are described in Section 7.

2   Background

A *partial truth assignment* to a set of Boolean variables V is a mapping $t : V \to \{0, 1, *\}$. For some variable $v \in V$, if $t(v) = 1$ then the literal v is said to be "true" while the literal $\bar{v}$ is said to "false" (and vice versa if $t(v) = 0$). If $t(v) = *$, v and $\bar{v}$ are said not to be assigned values. Let a *clause* denote a set of literals. A clause is *satisfied* by a truth assignment t iff at least one of its literals is true under t. A clause is said to be *violated* by a truth assignment t if all of its literals are false under t. A Boolean formula in conjunctive normal form (CNF) can be represented by a set C of clauses.

2.1   Clause Partitions

For a given Boolean formula in CNF, a partial truth assignment t is said to be *invalid* if it violates any clauses. Otherwise, it is *valid*. The implicit state representation used in the Cassatt algorithm relies on the partition of clauses implied by a given valid partial truth assignment. Each clause c must fall into exactly one of the three categories shown in Figure 1.

We are going to compactly represent multiple partial truth assignments that share the same set of assigned variables. Such partial truth assignments appear in the process of performing a BFS for satisfiability to a given depth. Note that simply knowing which variables have been assigned is enough to determine which clauses are *unassigned* or not *activated*. The actual truth values assigned differentiate *satisfied* clauses from *open* clauses. Therefore, if the assigned variables are known, we will store the set of *open* clauses rather than the actual partial truth assignment. While the latter may be impossible to recover, we will show in the following sections that the set of *open* clauses contains enough information to perform a BFS.

Another important observation is that only clauses which are "cut" by the vertical line in Figure 1 have the potential to be *open* clauses. The number of such cut clauses does not depend on the actual assigned values, but depends on the order in which variables

- **Unassigned Clauses**:
  Clauses whose literals are unassigned.
- **Satisfied Clauses**:
  Clauses which have at least one literal sat-
  isfied.
- **Open Clauses**:
  Clauses which have at least one, but not all
  of their literals assigned, and are not satis-
  fied.

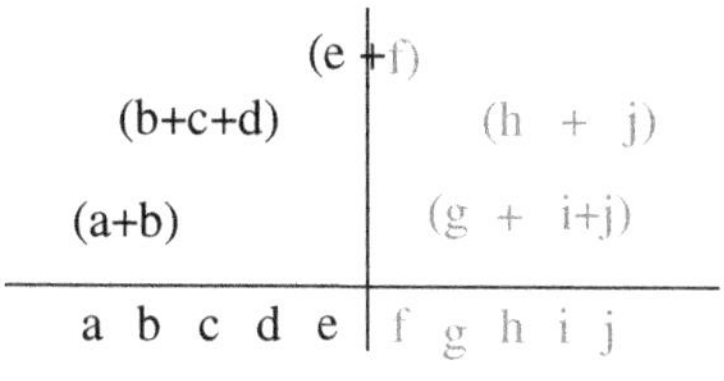

$$
\begin{array}{c|c}
 & (e+f) \\
(b+c+d) & (h+j) \\
(a+b) & (g+i+j) \\
\hline
a\ b\ c\ d\ e & f\ g\ h\ i\ j
\end{array}
$$

Fig. 1. The three clause partitions entailed by a valid partial truth assignment. The vertical line separates assigned variables (a through e) from unassigned variables (f through j, grayed out). We distinguish (i) clauses to the right of the vertical line, (ii) clauses to the left of the vertical line, and (iii) clauses cut by the vertical line.

are processed. This observation leads to the use of MINCE [1] as a heuristic for variable ordering since it reduces *cutwidth*. MINCE was originally proposed as a variable order-ing heuristic to complement nearly any SAT solver or BDD-based application. MINCE treats a given CNF as a (undirected) hypergraph and applies recursive balanced min-cut hypergraph bisection in order to produce an ordering of hypergraph vertices. This algo-rithm is empirically known to achieve small values of total hyperedge span. Intuitively, this algorithm tries to minimize cutwidth in many places, and as such is an ideal comple-ment to the Cassatt algorithm. Its runtime is estimated to be $O((V+C)log^2 V)$, where V is the number of variables and C is the number of clauses in a given CNF. By reducing the *cutwidth* in Cassatt, we obtain an exponential reduction in the number of possible sets of open clauses. However, storing each set explicitly is still often intractable. To efficiently store this collection of sets, we use ZDDs.

2.2 Zero-Suppressed Binary Decision Diagrams

In Cassatt we use ZDDs to attempt to represent a collection of N objects often in fewer than N bits. In the best case, N objects are stored in $O(p(logN))$ space where $p(n)$ is some polynomial. The compression in a BDD or ZDD comes from the fact that objects are represented by paths in a directed acyclic graph (DAG), which may have exponentially more paths than vertices or edges. Set operations are performed as graph traversals over the DAG. Zero-Suppressed Binary Decision Diagrams (ZDDs) are a variant of BDDs which are suited to storing collections of sets. An excellent tutorial on ZDDs is available at [14].

A ZDD is defined as a directed acyclic graph (DAG) where each node has a unique label, an integer index, and two outgoing edges which connect to what we will call *T-Child* and *E-child*. Because of this we can represent each node X as a 3-tuple $X\langle n, X_T, X_E\rangle$ where n is the index of the node X, X_T is its *T-Child*, and X_E is its *E-Child*. Each path in the DAG ends in one of two special nodes, the **0** node and the **1** node. These nodes have no successors. In addition, there is a single root node. When we use a ZDD we

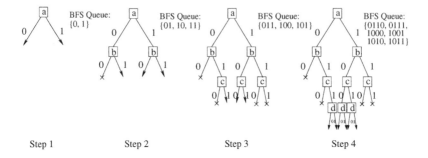

Fig. 2. Steps Taken by Breadth-First Search

will in reality keep a reference to the root node. The semantics of a ZDD can be defined recursively by defining the semantics of a given node.

A ZDD can be used to encode a collection of sets by encoding its characteristic function. We can evaluate a function represented by a ZDD by traversing the DAG beginning at the root node. At each node X, if the variable corresponding to the index of X is true, we select the *T-Child*. Otherwise we select with the *E-Child*. Eventually we will reach either **0** or **1**, indicating the value of the function on this input. We augment this with the *Zero-Suppression Rule*: we may eliminate nodes whose $T - Child$ is **0**. With these standard rules, **0** represents the empty collection of sets, while **1** represents the collection consisting of only the empty set. ZDDs interpreted this way have a standard set of operations based on recursive definitions [13], including the union and intersection of two collections of sets, for example.

3   A Motivating Example

$$\underbrace{(a+b)}_{1}\underbrace{(\bar{b}+c)}_{2}\underbrace{(d+e)}_{3}\underbrace{(\bar{a}+\bar{b}+\bar{c})}_{4}\underbrace{(c+\bar{d}+e)}_{5}$$

3.1   Steps Taken by Breadth-First Search

In general, BFS expands nodes in its queue until reaching a violated clause, at which point the search space is pruned. As a result, the number of nodes grows quickly.

The initial pass of the BFS considers the first variable a. Both $a = 0$ and $a = 1$ are valid partial truth assignments since neither violates any clauses. BFS continues by enqueueing both of these partial truth assignments. In Figure 2, the contents of the BFS Queue are listed as bit-strings. At Step 2, the BFS considers both possible values for b. Because of the clause $(a+b)$, BFS determines that $a = 0, b = 0$ is not a valid partial truth assignment. The remaining three partial truth assignments are valid, and BFS enqueues them. At Step 3, because of the clause $(\bar{a}+\bar{b}+\bar{c})$ we know that a, b, and c cannot all be true. The search space is pruned further because $(\bar{b}+c)$ removes all branches involving $b = 1, c = 0$. At Step 4, the state space doubles.

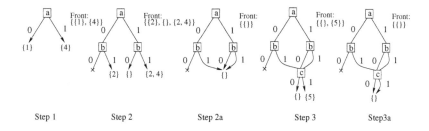

Fig. 3. Steps Taken by Cassatt

3.2 Steps Taken by Cassatt

The Cassatt algorithm performs a similar search, however partial truth assignments are not stored explicitly. As a result, if two partial truth assignments correspond to the same set of open clauses then only one of these will be stored. While the algorithm is described in Section 5, here we demonstrate possible reductions in the search space. We will call Cassatt's collection of sets of open clauses the *front*.

Variable a appears in two clauses: clause 1, $(a+b)$ and clause 4, $(\bar{a}+\bar{b}+\bar{c})$. If the variable a is assigned $a = 1$, then clause 1 becomes satisfied while making clause 4 *open*. If $a = 0$, then clause 1 becomes *open*. Because we cannot yet determine which of these assignments (if any) will lead to to a satisfying truth assignment, it is necessary to store both branches. The *front* shown in Figure 3, Step 1, is updated to contain the two sets of open clauses which correspond to the two possible valid partial truth assignments. Variable b appears in three clauses: clause 1, $(a+b)$, clause 2, $(\bar{b}+c)$, and clause 4, $(\bar{a}+\bar{b}+\bar{c})$. These clauses are our new set of activated clauses. For the first set, $\{1\}$, we consider both possible truth assignments, $b = 0$ and $b = 1$. Assigning $b = 0$ satisfies clauses 2 and 4. However, it does not satisfy clause 1. We also see that b corresponds to the end literal for clause 1. If we do not choose the value of b which satisfies this clause, it will never be satisfied. As a result we cannot include this branch in the new *front*. Note that realizing this only depends on noticing that b corresponds to the end literal for some open clause. For $b = 1$ we see that clause 1 is satisfied, but clauses 2 and 4 are not yet satisfied. So, $\{2,4\}$ should be in the *front* after this step is completed.

A similar analysis can be performed for the remaining set in the *front*, $\{4\}$. In Cassatt it is not necessary to consider these sets in succession; ZDD algorithms allow us to perform these operations on multiple sets simultaneously. These sets of clauses, listed in the *front* shown in Step 2, are all the state information Cassatt maintains. It is here that this algorithm begins to differ substantially from a naive BFS. One of the sets in the *front* is the empty set. This means that there is some assignment of variables which could satisfy every clause *in which any of these variables appear*. As a result, there is no need to examine any other truth assignment; the partial truth assignment which leaves no open clauses is clearly and provably the best choice. As a result of storing sets of open clauses, Cassatt prunes the search space and only considers the single node. Whereas the traditional BFS needs to consider three alternatives at this point, Cassatt, by storing open clauses, deduces immediately that one of these is provably superior. As a result, only a single node is expanded. As seen in Step 2a, this effectively restructures the search space so that several of the possible truth assignments are subsumed into a single branch.

As Cassatt processes the third variable, c, it needs to only consider its effects on the single, empty subset of open clauses. Since each new variable activates more clauses, this empty subset once again, can potentially grow. Again, one possible variable choice satisfies all activated clauses. This is shown in Step 3. As a result, Cassatt can once again subsume multiple truth assignments into a single branch as indicated in Step 3a.

4   Advantages of Storing Open Clauses

If two partial truth assignments create the exact same set of open clauses, then clearly space can be saved by only storing one of these subsets. This also cuts the search space. Moreover, if a stored set of clauses A is a proper subset of another set of clauses B, then the partial truth assignment corresponding to B is *sub-optimal* and can be discarded. This is so because *every* clause which is affected by any partial truth assignment thus far (every *activated* clause) has been considered when producing A and B. As a result, there are no consequences to choosing the truth assignment which produces A over the truth assignment which produces B. Every satisfying truth assignment based on the partial truth assignment giving B corresponds to a satisfying truth assignment based on A. These subsumption rules are naturally addressed by the ZDD data structure, which appears both appropriate and effective as a compact data structure.

We use three ZDD operations originally introduced in [4]. The *subsumed difference* X_A / X_B is defined as the collection of all sets in X_A which are not subsumed by some set in X_B. Based on the subsumed difference operator, it is possible to define an operator $NoSub(X)$ as the collection of all sets in X which are not subsumed by some other set in X. Finally, the *subsumption-free union* $X_A \bigcup_S X_B$ is defined as the collection of sets in $X_A \bigcup X_B$ from which all subsumed sets have been removed. For the sake of brevity, we will not repeat the full, recursive definitions of these operators here, but only point out that the operators introduced in [4] were shown for slightly different ZDD semantics — to encode collections of clauses. Here we use the original ZDD semantics proposed by Minato [13] — to encode collections of subsets. However, the definitions of operators themselves are unaffected. These operators provide a mechanism for maintaining a subsumption-free collection of sets.

ZDDs are known to often achieve good compression when storing sparse collections. In general a collection of sets may or may not compress well when represented by ZDDs. However in Cassatt our representation can be pruned by removing sets which are subsumed by another set. Intuitively, this leads to a sparser representation for two reasons: sets with small numbers of elements have a greater chance of subsuming larger sets in this collection, and each subsumption reduces the number of sets which need to be stored. By using operators which eliminate subsumed sets, then, we hope to improve the chance that our given collection will compress well. In Cassatt this corresponds to improved memory utilization and runtime.

A natural question to ask is whether the number of subsets to be stored (the *front*) has an upper bound. An easy bound based on the *cutwidth* is obtained by noting that for c clauses, the maximum number of incomparable subsets is exactly the size of the maximal anti-chain of the partially ordered set 2^c.

Theorem 4.1 [10, p. 444] The size of the maximal anti-chain of 2^c is given by $\left(\begin{array}{c} c \\ \lfloor \frac{c}{2} \rfloor \end{array} \right)$.

This simple upper bound does not take into account the details of Cassatt. Doing this may yield a substantially tighter bound.

It should be clear from the example that simply by storing sets of open clauses, Cassatt potentially searches a much smaller space than a traditional BFS. In general, it cannot fully avoid the explosion in space. Although there are potentially many more combinations of clauses than there are variables, Cassatt cannot examine more nodes than a straightforward BFS. At each stage, Cassatt has the potential to reduce the number of nodes searched. Even if no reduction is ever possible, Cassatt will only examine as many nodes as a traditional BFS.

5   Compressed Breadth-First Search

The Cassatt algorithm implements a breadth-first search based on keeping track of sets of open clauses from the given CNF formula. The collection of these sets represents the state of the algorithm. We refer to this collection of sets as the *front*. The algorithm advances this *front* just as a normal breadth-first algorithm searches successively deeper in the search space. The critical issue is how the *front* can be advanced in a breadth-first manner and determine satisfiability of a CNF formula. To explain this, we will first be precise about what should happen based on combining a single set of open clauses with a truth assignment to some variable v. We will then show how this can be done to the collection of sets by using standard operations on ZDDs.

Given a set of open clauses corresponding to a valid partial truth assignment t, and some assignment to a single variable $v \leftarrow r$, there are two main steps which must be explained. First, Cassatt determines if the combination of t with $v \leftarrow r$ produces a new valid partial truth assignment. Then, Cassatt must produce the new set of open clauses corresponding to the combination of t with $v \leftarrow r$. Since the *front* corresponds to the collection of valid partial truth assignments, then if this *front* is ever empty (equal to **0**), then the formula is unsatisfiable. If Cassatt processes all variables without the *front* becoming empty then no clauses will be activated, but unsatisfied. Thus if the formula is satisfiable, the *front* will contain only the empty set (equal to **1**) at the end of the search.

5.1   Detecting Violated Clauses

With respect to the variable ordering used, literals within a clause are divided into three categories: *beginning*, *middle*, and *end*. The *beginning* literal is processed first in the variable ordering while the *end* literal is processed last. The remaining literals are called *middle* literals. A nontrivial clause may have any (nonnegative) number of middle literals. However, it is useful to guarantee the existence of a *beginning* and an *end* literal. If a clause does not have at least 2 literals, then it forces the truth assignment of a variable, and the formula can be simplified. This is done as a preprocessing step in Cassatt. Cassatt knows in advance which literal is the *end* literal for a clause, and can use this information to detect violated clauses. Recall that a violated clause occurs when all the variables which appear within a clause have been assigned a truth value, yet the clause remains unsatisfied. Thus only the *end* literal has the potential to create a violated clause. The value which does not satisfy this clause produces a conflict or violated clause. This is the case since the sets stored in the *front* contain only unsatisfied clauses. After processing the *end* variable, no assignment to the remaining variables can affect satisfiability.

Determining if the combination of t and $v \leftarrow r$ is a valid partial truth assignment is thus equivalent to considering if v corresponds to an *end* literal for some clause c in the set corresponding to t. If the assignment $v \leftarrow r$ causes this literal to be $false$ then all literals in c are false, and combining t with $v \leftarrow r$ is not valid.

5.2   Determining the New Set of Open Clauses

Given a set of clauses S corresponding to some partial truth assignment and an assignment to a single variable v, all clauses in S which do not contain any literal corresponding to v's should, by default, propagate to the new set S'. This is true, since any assignment to v does not affect these clauses in any way. Consider a clause not in S containing v as its *beginning* variable. If the truth assignment given to v causes this clause to be satisfied, then it should *not* be added to S'. Only if the truth assignment causes the clause to remain unsatisfied does it become an open clause and should be added to S'. Finally if a clause in S contains v as one of its *middle* variables then the literal corresponding to this variable must again be examined. If the truth assignment given to v causes this clause to become satisfied, then it should not be propagated to S'. Otherwise, it should propagate from S to S'.

Table 1. New-Subset Rules

	Beginning	Middle	End	None
Satisfied	Impossible	No Action	No Action	No Action
Open	Impossible	**if**$(t(l) = 0)$ $S' \leftarrow S' \bigcup C$	No Action	$S' \leftarrow S' \bigcup C$
Unassigned	**if**$(t(l) = 0)$ $S' \leftarrow S' \bigcup C$	Impossible	Impossible	No Action

These rules are summarized in table 1. For a given clause and an assignment to some variable, the action to be taken can be determined by where the literal appears in the clause (column) and the current status of this clause (row). Here, "No Action" means that the clause should not be added to S'.

It should be noted that when an *end* variable appears in an open clause, no action is necessary only with regard to S' (it does not propagate into S'). If the assignment does not satisfy this clause, then the partial truth assignment is invalid. However, given that a partial truth assignment is valid then these clauses must become satisfied by that assignment.

5.3 Performing Multiple Operations via ZDDs

Cassatt's behavior for a single set was described in Table 1. However, the operations described here can be reformulated as operations on the entire collection of sets. By simply iterating over each set of open clauses and performing the operations in Table 1, one could achieve a *correct*, if inefficient implementation. However, it is possible to represent the combined effect on all sets in terms of ZDD operations. In order to create an *efficient* implementation, we perform these ZDD operations on the entire *front* rather than iterating over each set.

To illustrate this, consider how some of the clauses in each subset can be violated. Recall that if a clause c is violated, then the set containing c cannot lead to satisfiability and must be removed from the *front*. The same violated clauses will appear in many sets, and *each* set in the *front* containing any violated clause must be removed. Instead of iterating over each set, we intersect the collection of all *possible* sets without any violated clause with the *front*; all sets containing any violated clause will be removed. Such a collection can be formulated using ZDDs.

More completely, a truth assignment t to a single variable v has the following effects:

- It *violates* some clauses. Let $U_{v,t}$ be the set of all clauses c such that the *end* literal of c corresponds to v and is false under t. Then $U_{v,t}$ is the set of clauses which are violated by this variable assignment.
- It *satisfies* some clauses. Let $S_{v,t}$ be the set of all clauses c such that the literal of c corresponding to v is true under t. If these clauses were not yet satisfied, then they become satisfied by this assignment.
- It *opens* some clauses. Let $A_{v,t}$ be the set of all clauses c such that the *beginning* literal of c corresponds to v and is false under t. Then $A_{v,t}$ is the set of clauses which have just been activated, but remain unsatisfied.

Note that each of these sets depends only on the particular truth assignment to v. With each of these sets of clauses, an appropriate action can be taken on the entire *front*.

5.4 Updating Based on Newly Violated Clauses

Given a set of clauses $U_{v,t}$ which have just been violated by a truth assignment, all subsets which have one or more of these clauses must be removed from the *front*. Each

subset containing any of these clauses cannot yield satisfiability. We can build the collection of all sets which do
not contain any of these clauses. This new collection of
sets will have a very compact representation when using
ZDDs. Let $U'_{v,t}$ denote the collection of all possible sets of
clauses which do not contain any elements from $U_{v,t}$. That
is, $S \in U'_{v,t} \to S \bigcap U_{v,t} = \emptyset$. The ZDD corresponding to
$U'_{v,t}$ is shown in Figure 4.

In Figure 4, the dashed arrow represents a node's *E-Child* while a solid arrow represents a node's *T-Child*. Because of the Zero-Suppression rule, any node which does
not appear in the ZDD is understood not to appear in any
set in the collection. As a result, to create a ZDD without clauses in $U_{v,t}$, no element in $U_{v,t}$ should appear as a
node in the ZDD for $U'_{v,t}$. Because we look at a finite set
of clauses C, we wish to represent the *remaining clauses*
$C \setminus U_{v,t}$ as *don't care* nodes in the ZDD. To form the *don't
care* nodes, we simply create a ZDD node with both its
T-child and its *E-child* pointing to the same successor. The
resulting ZDD has the form shown in Figure 4a. When we

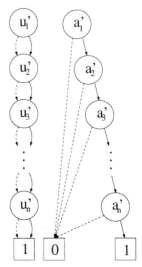

Fig. 4: (a) The $U'_{v,t}$ ZDD. (b)
The $A'_{v,t}$ ZDD

take *front* $\leftarrow$ *front* $\bigcap U'_{v,t}$ we will get all subsets which do not contain any clauses in
$U_{v,t}$. As a result, the *front* will be pruned of all partial truth which become invalid as a
result of this assignment.

Instead of including all clauses in $C \setminus U_{v,t}$, we can further reduce this operation by
including only clauses within the *cut*. Including nodes to preserve clauses outside the *cut*
is superfluous. The ZDD representing $U'_{v,t}$ can thus be created containing $O(cutwidth)$
nodes in the worst case.

5.5   Updating Based on Newly Satisfied Clauses

Consider a clause which has just been satisfied and appears somewhere in the *front*.
Since the *front* consists of sets of activated, unsatisfied clause, then it must have been
open (unsatisfied) under some partial truth assignment. However, this occurrence has just
been satisfied. As a result, *every* occurrence of a satisfied clause can simply be removed
entirely from the *front*.

---

for each $c \in S_{v,t}$ **do**
    $Temp_0 \leftarrow$ **Subset0**(Front, c)
    $Temp_1 \leftarrow$ **Subset1**(Front, c)
    Front $\leftarrow Temp_0 \bigcup Temp_1$

Fig. 5: Pseudocode for $\exists$ abstraction

Among other operations, the standard
operations on ZDDs include cofactoring,
or *Subset0* and *Subset1*. *Subset0(Z, a)*
creates the collection of all sets in the
ZDD Z without a given element a, while
Subset1(Z, a) creates the collection of
all sets in a ZDD Z which did contain a
given element a. However, the element a
is not present in any of the ZDDs returned by either *Subset0* or *Subset1*. As a result,
these operations can be used to eliminate a given element from the *front*. This idea is
illustrated in the pseudocode in Figure 5.

The combined effect of these operations on a ZDD is called *existential abstraction* [14] and can be implemented with a single operation on the ZDD [15].

5.6   Updating Newly Opened Clauses

Consider a set of newly opened clauses $A_{v,t}$. Each set in the *front* should now also include every element in $A_{v,t}$. We use the ZDD *product* to form the new *front* by letting $A'_{v,t} \equiv \{A_{v,t}\}$ and finding *front* $\times A'_{v,t}$.

When we build the ZDD $A'_{v,t}$, only those clauses in $A_{v,t}$ need to appear as nodes in $A'_{v,t}$. This is because of the Zero-Suppression rule: nodes not appearing in the ZDD $A'_{v,t}$ must be 0. The ZDD for $A'_{v,t}$ is simple as it only contains a single set. The form of all such ZDDs is shown in Figure 4b. When the ZDD product operation is used, *front* $\times A'_{v,t}$ will form the new *front*.

In many cases, a shortcut can be taken. By properly numbering clauses with respect to a given variable ordering, at each step we can ensure that each clause index is lower than the minimum index of the *front*. In this case, all clauses should be placed above the current *front*. The *front* will essentially remain unchanged, however new nodes in the form of Figure 4b will be added parenting the root node of the *front*. In this case, the overhead of this operation will be linear in the number of opened clauses.

> **CASSATT**($Vars, Clauses$)
>   Front $\leftarrow$ **1**
>   **for each** $v \in Vars$ **do**
>   Front† $\leftarrow$ Front
>   *Form sets* $U_{v,0}, U_{v,1}, S_{v,0}, S_{v,1}, A_{v,0}, A_{v,1}$
>   *Build the ZDDs* $U'_{v,0}, U'_{v,1}$
>   Front $\leftarrow$ Front $\bigcap U'_{v,1}$
>   Front† $\leftarrow$ Front† $\bigcap U'_{v,0}$
>   Front $\leftarrow$ $\exists$**Abstract**(Front, $S_{v,1}$)
>   Front† $\leftarrow$ $\exists$**Abstract**(Front†, $S_{v,0}$)
>   *Build the ZDDs* $A'_{v,0}, A'_{v,1}$
>   Front $\leftarrow$ Front $\times A'_{v,1}$
>   Front† $\leftarrow$ Front† $\times A'_{v,0}$
>   Front $\leftarrow$ Front $\bigcup_S$ Front†
>   **if** Front == **0 then**
>   **return** Unsatisfiable
>   **if** Front == **1 then**
>   **return** Satisfiable

Fig. 6: Pseudocode for the Cassatt Algorithm

5.7   Compressed BFS Pseudocode

Now that we have shown how the *front* can be modified to include the effects of a single variable assignment, the operation of the complete algorithm should be fairly clear. For each variable v, we will copy the *front* at a given step, then modify one copy to reflect assigning $v = 1$. We will modify the other copy to reflect assigning $v = 0$. The new *front* will be the union with subsumption of these two.

ZDD nodes are in reality managed via reference counts. When a ZDD (or some of its nodes) are no longer needed, we decrement this reference count. When we copy a ZDD, we simply increment the reference count of the root node. Thus, copying the *front* in

Cassatt takes only constant overhead. The pseudocode of the Cassatt algorithm is shown in Figure 6.

Here, we order the steps of the algorithm according to the MINCE ordering, applied as a separate step before Cassatt begins. We initially set the *front* to the collection containing the empty set. This is consistent, because trivially there are no open clauses yet. Recall that after all variables are processed, either the *front* will contain only the empty set (be **1**), or will be the empty collection of sets (be **0**).

6   Empirical Validation

We implemented Cassatt in C++ using the CUDD package [18]. We also used an *existential abstraction* routine from the Extra library [15]. For these results, we disabled reordering and garbage collection. These tests were performed on an AMD Athlon 1.2GHz machine, with 1024MB of 200MHz DDR RAM running Debian Linux. We also include performance of Chaff [16] and GRASP [17], two leading DLL-based SAT solvers. We also include results of ZRes, a solver which performs the DP procedure [7] while compressing clauses with a ZDD. All solvers were used with their default configurations, except ZRes which requires a special switch when solving hole-n instances. All runs were set to time out after 500s.

We ran two sets of benchmarks which are considered to be difficult for traditional SAT solvers [3]. The *hole-n* family of benchmarks are come from the pigeonhole principle. Given $n+1$ pigeons and n holes, the pigeonhole principle implies that least 2 pigeons must live in the same hole. The *hole-n* benchmarks are a CNF encoding of the negation of this principle, and are unsatisfiable as they place at most 1 pigeon in each hole. The number of clauses in this family of benchmarks grows as $\Theta(n^3)$ while the number of variables grows as $\Theta(n^2)$. For hole-50, there are 2550 variables and 63801 clauses. The *Urquhart* benchmarks are a randomly generated family relating to a class of problems based on expander graphs [20]. Both families have been used to prove lower bounds for runtime of classical DP and DLL algorithms and contain only unsatisfiable instances. [3] shows that any DLL or resolution procedure requires $\Omega(2^{n/20})$ time for hole-n instances. Figure 7 empirically demonstrates that Cassatt requires $\Theta(n^4)$ time for these instances. This does not include time used to generate the MINCE variable ordering or I/O time.

The runtimes of the four solvers tested on instances from hole-n are graphed in Figure 7a. The smallest instance we consider is hole-5; Cassatt efficiently proves unsatisfiability of hole-50 in under 14 seconds. Figure 7b plots the runtimes of Cassatt on hole-n vs. n for $n < 50$ on a log-log scale. We also include a plot of cn^4 with a constant set to clearly show the relationship with the empirical data (on the log-log scale, this effectively translates the plot vertically).

The Urquhart-n instances have some degree of randomness. To help compensate for this, we tested 10 different randomly generated instances for each n, and took the average runtime of those instances which completed. Memory appears to be the limiting factor for Cassatt, rather than runtime; it cannot solve one Urq-8 instance, while quickly solving others. With respect to the DIMACS benchmark suite, Cassatt efficiently solves many families. Most *aim* benchmarks are solved, as well as *pret*, and *dubois* instances. However, DLL solvers outperform Cassatt on many benchmarks in this suite.

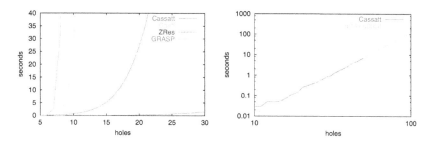

Fig. 7. (a) Runtime of four SAT solvers on hole-n. (b) Cassatt's runtime on hole-n. Cassatt achieves asymptotic speedup over three other solvers

Table 2. Benchmark times and completion ratios for Urquhart-n. Timeouts were set to 500s

Urq-n	Average		Cassatt		Chaff [16]		ZRes [5]		Grasp [17]	
	# Vars	# Clauses	% Solved	Avg	% Solved	Avg	% Solved	Avg	% Solved	Avg
3	44.6	421.2	100%	.07s	50%	139s	100%	.20s	0%	-
4	80.2	781.4	100%	.39s	0%	-	100%	.57s	0%	-
5	124.4	1214.4	100%	1.32s	0%	-	100%	1.35s	0%	-
6	177.4	1633.6	100%	4.33s	0%	-	100%	2.83s	0%	-
7	241.8	2261.4	100%	43.27s	0%	-	100%	5.39s	0%	-
8	315.2	2936.6	90%	43.40s	0%	-	100%	9.20s	0%	-

7   Conclusions

In this paper, we have introduced the Cassatt algorithm, a compressed BFS which solves the SAT decision problem. It leverages the compression power of ZDDs to compact state representation. Also, the novel method for encoding the state uses subsuming semantics of ZDDs to reduce redundancies in the search. In addition, the runtime of this algorithm is dependent on the size of the compressed, implicit representation rather than on the size of an explicit representation. This is the first work to attempt using a compressed BFS as a method for solving the SAT decision problem. Our empirical results show that Cassatt is able to outperform DP and DLL based solvers in some cases. We show this by examining runtimes for Cassatt and comparing this to proven lower bounds for DP and DLL based procedures.

Our ongoing work aims to add Boolean Constraint Propagation to the Cassatt algorithm. With Constraint Propagation, any unsatisfied clause which has only one remaining unassigned variable forces the assignment of that variable. It is hoped that taking this into account will give a reduction in memory requirements and runtime. Another direction for future research is studying the effects of SAT variable and BDD orderings on the performance of the proposed algorithm.

References

1. F. A. Aloul, I. L. Markov, and K. A. Sakallah. Faster SAT and Smaller BDDs via Common Function Structure. *Proc. Intl. Conf. Computer-Aided Design*, 2001.
2. F. A. Aloul, M. Mneimneh, and K. A. Sakallah. Backtrack Search Using ZBDDs. *Intl. Workshop on Logic and Synthesis, (IWLS)*, 2001.
3. P. Beame and R. Karp. The efficiency of resolution and Davis-Putnam procedures . submitted for publication.
4. P. Chatalic and L. Simon. Multi-Resolution on Compressed Sets of Clauses. *Proc. of 12th International Conference on Tools with Artificial Intelligence (ICTAI-2000)*, November 2000.
5. P. Chatalic and L. Simon. ZRes: the old DP meets ZBDDs. *Proc. of the 17th Conf. of Autom. Deduction (CADE)*, 2000.
6. M. Davis, G. Logemann, and D. Loveland. A Machine Program for Theorem Proving. *Comm. ACM*, 5:394–397, 1962.
7. M. Davis and H. Putnam. A computing procedure for quantification theory. *Jounal of the ACM*, 7:201–215, 1960.
8. P. Ferragina and G. Manzini. An experimental study of a compressed index. *Proc. 12th ACM-SIAM Symposium on Discrete Algorithms (SODA)*, 2001.
9. R. Gasser. *Harnessing Computational Resources for Efficient Exhaustive Search*. PhD thesis, Swiss Fed Inst Tech, Zurich, 1994.
10. R. L. Graham, M. Grötschel, and L. Lovász, editors. *Handbook of Combinatorics*. MIT Press, January 1996.
11. J. F. Groote and H. Zantema. Resolution and binary decision diagrams cannot simulate each other polynomially. Technical Report UU-CS-2000-14, Utrecht University, 2000.
12. A. San Miguel. Random 3-SAT and BDDs: The Plot Thickens Further . *CP*, 2001.
13. S. Minato. Zero-Suppressed BDDs for Set Manipulation in Combinatorial Problems. *30th ACM/IEEE DAC*, 1993.
14. A. Mishchenko. An Introduction to Zero-Suppressed Binary Decision Diagrams. http://www.ee.pdx.edu/~alanmi/research/.
15. A. Mishchenko. EXTRA v. 1.3: Software Library Extending CUDD Package: Release 2.3.x. http://www.ee.pdx.edu/~alanmi/research/extra.htm.
16. M. Moskewicz et al. Chaff: Engineering an Efficient SAT Solver . *Proc. of IEEE/ACM DAC*, pages 530–535, 2001.
17. J. P. Marques Silva and K. A. Sakallah. GRASP: A new search algorithm for satisfiability. *ICCAD* , 1996.
18. F. Somenzi. CUDD: CU Decision Diagram Package Release 2.3.1. http://vlsi.colorado.edu/~fabio/CUDD/cuddIntro.html.
19. T. E. Uribe and M. E. Stickel. Ordered binary decision diagrams and the Davis-Putnam procedure. In J. P. Jouannaud, editor, *1st Intl. Conf. on Constraints in Comp. Logics*, volume 845 of *LNCS*, pages 34–49. Springer, September 1994.
20. A. Urquhart. Hard examples for resolution. *Journal of the ACM*, 34, 1987.

Using Multi-level Graphs for Timetable Information in Railway Systems[*]

Frank Schulz[1,2], Dorothea Wagner[1], and Christos Zaroliagis[2]

[1] Department of Computer and Information Science
University of Konstanz
Box D188, 78457 Konstanz, Germany
{Frank.Schulz, Dorothea.Wagner}@uni-konstanz.de

[2] Computer Technology Institute, and
Department of Computer Engineering & Informatics
University of Patras, 26500 Patras, Greece
zaro@ceid.upatras.gr

Abstract. In many fields of application, shortest path finding problems in very large graphs arise. Scenarios where large numbers of on-line queries for shortest paths have to be processed in real-time appear for example in traffic information systems. In such systems, the techniques considered to speed up the shortest path computation are usually based on precomputed information. One approach proposed often in this context is a space reduction, where precomputed shortest paths are replaced by single edges with weight equal to the length of the corresponding shortest path. In this paper, we give a first systematic experimental study of such a space reduction approach. We introduce the concept of multi-level graph decomposition. For one specific application scenario from the field of timetable information in public transport, we perform a detailed analysis and experimental evaluation of shortest path computations based on multi-level graph decomposition.

1 Introduction

In this paper we consider a scenario where a large number of on-line shortest path queries in a huge graph has to be processed as fast as possible. This scenario arises in many practical applications, including route planning for car traffic [11,4,12, 17,18,13], database queries [16], Web searching [2], and time-table information in public transport [20,3,10]. The algorithmic core problem consists in performing Dijkstra's shortest path algorithm using appropriate speed-up techniques.

Our initial interest in the problem stems from our previous work on time table information in railway systems [20]. In such a problem, the system has to answer on-line a potentially infinite number of customer queries for optimal (e.g.,

---

[*] This work was partially supported by the Human Potential Programme of EU under contract no. HPRN-CT-1999-00104 (AMORE), by the Future and Emerging Technologies Programme of EU under contract no. IST-1999-14186 (ALCOM-FT), and by the Deutsche Forschungsgemeinschaft under grant WA 654/12-1.

D. Mount and C. Stein (Eds.): ALENEX 2002, LNCS 2409, pp. 43–59, 2002.
© Springer-Verlag Berlin Heidelberg 2002

fastest) travel connections in a wide-area network. The concrete scenario comes from the Hafas central server [9] of the German railways: the server is directly accessible to any customer either through terminals in the train stations, or through a web interface. Note that space consumption is not the major issue in such a scenario. What it matters most is the average (as opposed to the maximum) response time for a query.

In practice, the usual approach to tackle the shortest path problems arising in scenarios like the above is to use heuristic methods, which in turn implies that there is no guarantee for an optimal answer. On the contrary, we are interested in *distance-preserving* algorithms, i.e., shortest path algorithms that produce an optimal answer for any input instance. Distance-preserving algorithms were not in wide use in traffic information systems, mainly because the average response time was perceived to be unacceptable. However, the results in [20,3] showed that distance-preserving variants of Dijkstra's algorithm are competitive in the sense that they do not constitute the bottleneck operation in the above scenario. These are the only publications known to us that investigate distance-preserving speed-up techniques of Dijkstra's algorithm. The recent work in [10] investigates multi-criteria shortest path problems for computing Pareto optimal solutions in the above scenario. All these publications [20,3,10] are the only ones known to us regarding algorithms for wide-area railway traffic information systems. Related work is known for other traffic engineering systems, concerning mainly local public transport [15], or private transport in wide-area networks [11,1,4,17, 12,13,15,18]. For various reasons (see e.g., [20]) the techniques in those papers cannot be directly applied to wide-area railway traffic information systems.

Several of the approaches used so far in traffic engineering introduce speed-up techniques based on hierarchical decomposition. For example, in [11,1,4,13] graph models are defined to abstract and store road maps for various routing planners for private transport. Similarly, in [19] a space reduction method for shortest paths in a transportation network is introduced. The idea behind such techniques is to reduce the size of the graph in which shortest path queries are processed by replacing precomputed shortest paths by edges. The techniques are hierarchical in the sense that the decomposition may be repeated recursively. Several theoretical results on shortest paths, regarding planar graphs [7,8,14] and graphs of small treewidth [6,5], are based on the same intuition.

So far, however, there exists no systematic evaluation of hierarchical decomposition techniques, especially when concrete application scenarios are considered. In [20], a first attempt is made to introduce and evaluate a speed-up technique based on hierarchical decomposition, called selection of stations. Based on a small set of selected vertices an auxiliary graph is constructed, where edges between selected vertices correspond to shortest paths in the original graph. Consequently, shortest path queries can be processed by performing parts of the shortest path computation in the much smaller and sparser auxiliary graph. In [20], this approach is extensively studied for one single choice of selected vertices, and the results are quite promising.

In this paper, we follow up and focus on a detailed and systematic experimental study of such a hierarchical decomposition technique. We introduce the *multi-level graph model* that generalizes the approach of [20]. A multi-level graph $\mathcal{M}$ of a given weighted digraph $G = (V, E)$ is a digraph which is determined by a sequence of subsets of V and which extends E by adding multiple levels of edges. This allows to efficiently construct a subgraph of $\mathcal{M}$ which is substantially smaller than G and in which the shortest path distance between any of its vertices is equal to the shortest path distance between the same vertices in G. Under the new framework, the auxiliary graph used in [20] – based on the selection of stations – can be viewed as adding just one level of edges to the original graph.

We implemented and evaluated a distance-preserving speed-up technique based on a hierarchical decomposition using the multi-level graph model. Our study is based on all train data (winter period 1996/97) of the German railways consisting of time-table information and queries. The processed queries are a snapshot of the central Hafas server in which all queries of customers of all ticket offices in Germany were recorded over several hours. From the time-table information, the so-called *train graph* is generated in a preprocessing step. Answering a connection query corresponds in solving a shortest path in the train graph. Based on that graph, we considered various numbers l of levels and sequences of subsets of vertices. For each of these values, the corresponding multi-level graphs are evaluated. Our study was concentrated in measuring the improvement in the performance of Dijkstra's algorithm when it is applied to a subgraph of $\mathcal{M}$ instead of being applied to the original train graph. Our experiments demonstrated a clear speed-up of the hierarchical decomposition approach based on multi-level graphs. More precisely, we first considered various selection criteria for including vertices on the subsets which determine the multi-level graphs. This investigation revealed that random selection (as e.g., proposed in [21]) is a very bad choice. After choosing the best criteria for including vertices in the subsets, we analyzed their sizes and demonstrated the best values for these sizes. It turns out that the dependence of the multi-level graphs on the subset sizes is also crucial. Finally, for the best choices of subsets and their sizes, we determined the best values for the number of levels. For the best choice of all parameters considered we obtained a speed-up of about 11 for CPU time and of about 17 for the number of edges hit by Dijkstra's algorithm.

2   Multi-level Graph

Let $G = (V, E)$ be a weighted digraph with non-negative edge weights. The *length* of a path is the sum of the weights of the edges in the path. The *multi-level* graph $\mathcal{M}$ of G is, roughly speaking, a graph that extends G in two ways:

1. It extends the edge-set of G by multiple *levels* of edges.
2. It provides the functionality to determine for a pair of vertices $s, t \in V$ a subgraph of $\mathcal{M}$ such that the length of a shortest path from s to t in that subgraph is equal to the shortest path length in G. To achieve this, we

use a special data structure called the *component tree* (a tree of connected components).

The objective is the resulting subgraph of $\mathcal{M}$ to be substantially smaller than the original graph G. Then, single-pair shortest path algorithms can be applied to the smaller graph, improving the performance.

The multi-level graph is built on the following input:

- a weighted digraph $G = (V, E)$ consisting of vertices V and edges $E \subseteq V \times V$
- a sequence of l subsets of vertices S_i $(1 \leq i \leq l)$, which are decreasing with respect to set inclusion: $V \supset S_1 \supset S_2 \supset \ldots \supset S_l$

To emphasize the dependence on G and the sets $S_1, \ldots, S_l$, we shall refer to the multi-level graph by $\mathcal{M}(G; S_1, \ldots, S_l)$. The vertex-sets S_i will determine the levels of the multi-level graph. In the following, we shall discuss the construction of the multi-level graph and of the component tree.

2.1   Levels

Each level of $\mathcal{M}(G; S_1, \ldots, S_l)$ is determined by a set of edges. The endpoints of these edges determine the vertex set of each level. For each set S_i $(1 \leq i \leq l)$, we construct three sets of edges:

- **level edges**: $E_i \subseteq S_i \times S_i$
- **upward edges**: $U_i \subseteq (S_{i-1} \setminus S_i) \times S_i$
- **downward edges**: $D_i \subseteq S_i \times (S_{i-1} \setminus S_i)$

We call the triple $L_i := (E_i, U_i, D_i)$ the **level** i of the multi-level graph. We further say that $L_0 := (E, \emptyset, \emptyset)$ is the **level zero**, where E are the edges of the original graph G. With the level zero there are totally $l+1$ levels, so we say that $\mathcal{M}(G; S_1, \ldots, S_l)$ is an $l + 1$-level graph. Figure 1 illustrates a 3-level graph.

Construction. The construction of the levels is iterative, so we assume that we have already constructed the level L_{i-1}. The iteration begins with $i = 1$. For each vertex u in S_{i-1} consider a shortest-path tree T_u (rooted at u) in the graph (S_{i-1}, E_{i-1}). Candidates for edges in level L_i are all the edges $S_i \times S_i$ for level edges, $(S_{i-1} \setminus S_i) \times S_i$ for upward edges, and $S_i \times (S_{i-1} \setminus S_i)$ for downward edges. The condition to decide whether one candidate edge (u, v) is actually taken for the sets E_i, U_i and D_i is the following:

L_i contains an edge (u, v) if and only if no internal vertex of the u-v path in T_u belongs to S_i.

In other words, if the u-v path contains no vertex of S_i except for the two endpoints u and v, the edge (u, v) is added to L_i. The weight of a new edge (u, v) is the shortest path length from u to v in G.

Note that the level L_i is not uniquely determined by this construction, since the shortest-path trees are not unique. Now, we can define the multi-level graph as

$$\mathcal{M}(G; S_1, \ldots, S_l) := (V, E \cup \bigcup_{i=1\ldots l} (E_i \cup U_i \cup D_i))$$

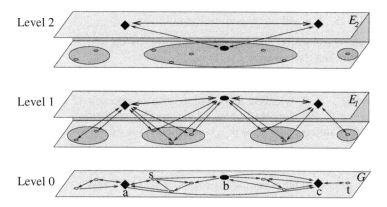

Fig. 1. A simple example of a 3-level graph. Level zero consists of the original graph G. The sets of vertices that define the 3-level graph are $S_1 = \{a, b, c\}$ and $S_2 = \{a, c\}$. In order to show the levels, we draw copies of each vertex for the levels one and two, but actually there is only one occurrence of them in the 3-level graph. The levels one and two are each split into two planes, where the upper plane contains the edges E_i, and the lower plane shows the connected components in the graph $G - S_i$. The edges U_i and D_i connect vertices in different planes of one level.

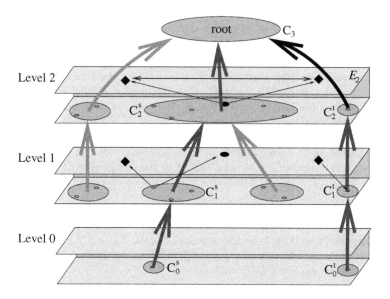

Fig. 2. The component tree for the 3-level graph in Figure 1. Only the leaves for the vertices s and t are shown. The thin black edges are the edges E_{st} that define the subgraph with the same shortest path length as G.

Connected Components. Consider the subgraph of G that is induced by the vertices $V \setminus S_i$. We will use the following notation:

- the set of connected components is denoted by $\mathcal{C}_i$, and a single component is usually referred to by C;
- $V(C)$ denotes the set of vertices of a connected component C of $\mathcal{C}_i$;
- for a vertex $v \in V \setminus S_i$, let C_i^v denote the component in $\mathcal{C}_i$ that contains v;
- a vertex $v \in S_i$ is called **adjacent** to the component $C \in \mathcal{C}_i$, if v and a vertex of C are connected by an edge (ignoring direction);
- the set of adjacent vertices of a component C is denoted by $Adj(C)$.

The edges E_i, U_i and D_i can be interpreted in terms of connected components as follows (see Figure 1). The edges E_i resemble the shortest paths between vertices of S_i that pass through a connected component, i.e., if two vertices x and y are adjacent to the same component, and the shortest path from x to y is inside that component, then there is an edge from x to y representing that shortest path. This includes edges in G that connect two vertices of S_i. Notice that for a pair of vertices in S_i, the subgraph of $\mathcal{M}$ induced by E_i suffices to compute a shortest path between these vertices.

In the same way, the edges U_i represent shortest paths from a vertex inside a connected component to all vertices of S_i adjacent to that component, and the edges D_i represent the shortest paths from the adjacent vertices of a component to a vertex of the component.

2.2   Component Tree

The data structure to determine the subgraph of $\mathcal{M}$ for a pair of vertices $s, t \in V$ is a tree with the components $\mathcal{C}_1 \cup \ldots \cup \mathcal{C}_l$ as nodes. Additionally, there is a root C_{l+1}, and for every vertex $v \in V$ a leaf C_0^v in the tree (we assume that $Adj(C_0^v) := \{v\}$ and $Adj(C_{l+1}) := \emptyset$). The parent of a leaf C_0^v is determined as follows: Let i be the largest i with $v \in S_i$. If $i = l$, the parent is the root C_{l+1}. Otherwise, the smallest level where v is contained in a connected component is level $i + 1$, and the parent of C_0^v is the component $C_{i+1}^v \in \mathcal{C}_{i+1}$.

The parent of the components in $\mathcal{C}_l$ is also the root C_{l+1}. For one of the remaining components $C_i \in \mathcal{C}_i$, the parent is the component $C'_{i+1} \in \mathcal{C}_{i+1}$ with $V(C_i) \subseteq V(C'_{i+1})$. Figure 2 illustrates the component tree of the 3-level graph in Figure 1.

Subgraph. For the given pair of vertices $s, t \in V$ we consider the C_0^s-C_0^t path in the component tree. Let L be the smallest L with $C_L^s = C_L^t$ (i.e., $C_L^s = C_L^t$ is the lowest common ancestor of C_0^s and C_0^t in the tree). Then, with our notation for the components, the C_0^s-C_0^t path is

$$(C_0^s, C_k^s, C_{k+1}^s, \ldots, C_L^s = C_L^t, \ldots, C_{k'+1}^t, C_{k'}^t, C_0^t)$$

where $k > 0$ and $k' > 0$ are the levels of the parents of C_0^s and C_0^t as defined above (cf. darker tree edges in Figure 2). The subgraph with the same s-t shortest-path length as G is the subgraph $\mathcal{M}_{st}$ of $\mathcal{M}$ induced by the following edge set:

$$E_{st} := E_{L-1}$$
$$\cup \bigcup_{i=k,\ldots,L-1} \{(u,v) \in U_i | u \in Adj(C^s_{i-1}), v \in Adj(C^s_i)\}$$
$$\cup \bigcup_{i=k',\ldots,L-1} \{(u,v) \in D_i | u \in Adj(C^t_i), v \in Adj(C^t_{i-1})\}$$

The following lemma holds for shortest paths in $\mathcal{M}_{st}$.

Lemma 1. *The length of a shortest s-t path is the same in the graphs G and* $\mathcal{M}_{st}(G; S_1, \ldots, S_l)$.

Proof. [sketch] Let $s, t \in V$ be a pair of vertices G for which a s-t path in G exists, and let $C^s_0, C^s_k, \ldots, C^s_L = C^t_L, \ldots, C^t_{k'}, C^t_0$ be the corresponding graph in the component tree. By definition, every edge (u,v) in $\mathcal{M}_{st}$ has a weight that is at least as large as the shortest-path length from u to v in G. Hence, the length of a shortest s-t path in $\mathcal{M}_{st}$ can never be smaller than the one in G. It remains to prove that there is a s-t path in $\mathcal{M}_{st}$ with the same length as a shortest s-t path P in G. To prove this, it suffices to prove the following claims, where $1 \leq x \leq l$:

1. For each pair of vertices $u, v \in S_x$ such that there exists a u-v path in G, the graph (S_x, E_x) contains a path with the same length as a shortest u-v path in G.
2. For the subgraph $\mathcal{M}'$ of $\mathcal{M}$ induced by the edge set

$$E_x \cup \bigcup_{i=k,\ldots,x} \{(u,v) \in U_i | u \in Adj(C^s_{i-1}), v \in Adj(C^s_i)\}$$

   it holds that for each vertex $w \in S_x$ that is reachable from s in G there exists a path from s to w in $\mathcal{M}'$ with the same length as a shortest s-w path in G.
3. For the subgraph $\mathcal{M}'$ of $\mathcal{M}$ induced by the edge set

$$E_x \cup \bigcup_{i=k',\ldots,x} \{(u,v) \in D_i | u \in Adj(C^t_i), v \in Adj(C^t_{i-1})\}$$

   it holds that for each vertex $w \in S_x$ from which t is reachable in G there exists a path from w to t in $\mathcal{M}'$ with the same length as a shortest w-t path in G.

We first show how the proof is completed using the above claims, and then give the proofs of the claims. The value L is the level of the lowest common ancestor of C^s_0 and C^t_0 in the component tree. Because of this, s and t are in different components of the subgraph induced by $V - S_{L-1}$, and therefore at least one vertex of a shortest s-t path in G has to be in S_{L-1}. Let w (resp. z) be the first (resp. last) vertex of P that belongs to S_{L-1}. Then, vertices w and z

split P into three (not necessarily non-empty) parts P_1, P_2 and P_3. By Claim 1, it follows that there is a w-z path in $\mathcal{M}_{st}$ with the same length as P_2. Similarly, by Claim 2, it follows that there is a path in $\mathcal{M}_{st}$ from s to w with the same length as P_1, and by Claim 3 that there is a path in $\mathcal{M}_{st}$ from z to t with the same length as P_3. The concatenation of these three paths is an s-t path in $\mathcal{M}_{st}$ with the same length as P.

We now turn to the proofs of the claims. The proofs are by induction on x. We give the proof of Claim 1; the proofs of the other claims follow similarly. We start with the basis of the induction ($x = 1$).

Let u and v be two vertices of S_1 and $P = (u = v_1, \ldots, v_z = v)$ be the shortest u-v path in the shortest-path tree T_u in G considered in the definition of the levels. If no internal vertex of that path belongs to S_1, by the definition of E_1, there is an edge $(u, v) \in E_1$ whose weight is the length of P, and we are done. Otherwise, some of the internal vertices of P belong to S_1, and we consider all the subpaths P_j of P, where P_1 is the part from u to the first vertex belonging to S_1, then P_2 is the part from the latter vertex to the second vertex in P belonging to S_1, and so on. The end-vertices of each subpath P_j are connected by an edge in E_1, because for these subpaths there is no internal vertex in S_1, and the weight of such an edge is exactly the length of P_j in G. The combination of all these edges is the path in (S_1, E_1) we are looking for.

Now, assume that the claim is true for any value smaller than x. Then, the induction step for x is proved in exactly the same way as for the basis, by replacing G by (S_{x-1}, E_{x-1}), S_1 by S_x, and E_1 by E_x.   ∎

3   Graphs for Timetable Information

In the following the graphs used for timetable information will be defined, and some customizations of the multi-level approach needed for timetable information graphs will be discussed.

3.1   Train and Station Graph

The timetable information system that we consider is based on a timetable for a set of trains. For the sake of simplification we assume that the timetable is periodic with a period of one day, and that the objective of the system is to provide a train connection with earliest arrival time. A query consists of a departure station, an arrival station, and a departure time. That problem is reduced to a shortest-path problem in the train graph.

Train Graph TG. It contains a *vertex* v for every arrival and departure of a train, and two kinds of edges:

- *stay-edges*: for each station, all arrivals and departures v_i of trains are sorted according to the time the trains leave or arrive at the station, say $v_1, \ldots, v_n$. Then, the edges (v_i, v_{i+1}) for $i = 1 \ldots n - 1$ and (v_n, v_1) model *stays* at that station, where the last edge models a stay over midnight. The *weight* of a stay-edge is the duration of the stay in minutes.

- *travel-edges*: for each departure of a train, there is an edge from that departure to the arrival at the station where the train stops next. The *weight* of a travel-edge is the time difference in minutes between the arrival and the departure.

It is easy to see that solving a query amounts to computing a shortest path in the train graph from the departure vertex at the departure station (determined by the departure time) to one of the vertices of the arrival station. Note that this is a kind of a *single-source some-target* shortest path problem, where the targets for one query are the set of vertices belonging to one station. Because of this, we will need a second graph, the station graph.

Station Graph SG. It contains one vertex per railway station R, and there is an edge between two stations R_1 and R_2 if and only if there is an edge (v_1, v_2) in the train graph, with v_1 belonging to station R_1 and v_2 belonging to R_2. The station graph is simple and unweighted. With $T(R)$ we denote the set of all arrival and departure vertices in the train graph that belong to the station R. Note that the station graph is the graph minor of the train graph obtained by contracting all stay-edges in the train graph and by removing all but one of multiple edges. The following lemma follows directly by the definition of SG and TG.

Lemma 2. *Consider a subset Σ of vertices in SG, and let $T(\Sigma)$ be the set of all arrivals and departures of the stations in Σ. Then, if the stations $R_1, \ldots, R_k$ belong to one connected component of $SG - \Sigma$, the vertices $T(R_1), \ldots, T(R_k)$ belong to one connected component of $TG - T(\Sigma)$, and vice versa.*

3.2 Customization of the Multi-level Graph Model

If we define the multi-level graph $\mathcal{M}(TG)$ of TG according to the definition given in Section 2, then we would get a subgraph of $\mathcal{M}(TG)$ for a pair s, t of vertices on which we could solve a single-pair shortest path problem in order to determine an s-t shortest path in TG. In our case, however, we have to solve a single-source *some*-targets problem, and hence this is not actually suitable for our case. Instead, we need a subgraph that guarantees the same shortest-path length between every pair of vertices belonging to two *stations* (i.e., sets of vertices of TG). Therefore, we define on TG a slightly modified version of a multi-level graph:

1. The first modification is to start with a sequence of l sets of stations of the station graph, Σ_i ($1 \leq i \leq l$), which are decreasing with respect to set inclusion. Then, the l sets of vertices of the train graph are defined to be $S_i := \cup_{R \in \Sigma_i} T(R)$, all departures and arrivals of all the stations in Σ_i. The levels of the multi-level graph $\mathcal{M}$ are then defined using the S_i as described in Section 2.1 (page 46), yielding $\mathcal{M}(TG; S_1, \ldots, S_l)$. To emphasize the dependence of S_i on Σ_i and in order to facilitate notation, we shall refer to this multi-level graph as $\mathcal{M}(TG; \Sigma_1, \ldots, \Sigma_l)$.

2. The component tree is computed in the station graph. There is one leaf C^R per station R, and $Adj(C^R) := T(R)$, i.e., the arrivals and departures belonging to R.
3. We define a vertex v of the train graph to be adjacent to a component C of the station graph, if v and any vertex belonging to a station of C are connected by an edge in the train graph. With this definition, and s and t being the departure and arrival *stations*, the definition of the subgraph $\mathcal{M}_{st}$ is exactly the same as for general multi-level graphs (see Section 2.2 on page 48).

Given a query with departure station s, arrival station t, and a departure time, the subgraph $\mathcal{M}_{st}$ of $\mathcal{M}(TG; \Sigma_1, \ldots, \Sigma_l)$ depends now on the *stations* s and t. The departure time determines the departure vertex in TG belonging to station s. To solve the query, we have to compute the shortest-path length from the departure vertex of TG to one of the vertices belonging to station t. Based on Lemmata 1 and 2, we are able to show (next lemma) that it is sufficient to perform such a shortest path computation in $\mathcal{M}_{st}$.

Lemma 3. *For each departure vertex v in the train graph belonging to station s, the shortest-path length from v to one of the vertices belonging to station t is the same in the graphs TG and $\mathcal{M}_{st}(TG; \Sigma_1, \ldots, \Sigma_l)$.*

Proof. [sketch] Using Lemma 2 on page 51, the proof of Lemma 1 can be adopted to the customizations that were made for the train graph.

The proof for Claim 1 is exactly the same here. Claims 2 and 3 are modified in the way that now s and t are sets of vertices of TG, namely the sets of all arrivals and departures belonging to the stations s and t, respectively. Then, Claims 2 and 3 hold for each of these vertices, because of Lemma 2.

Let P be a shortest path in TG from the departure vertex v to one of the vertices belonging to station t. Then, similarly to the proof of Lemma 1, we can show that there is a path with the same end-vertices and of the same length in $\mathcal{M}_{st}(TG; \Sigma_1, \ldots, \Sigma_l)$.   ■

4   Experiments

As mentioned in the introduction, we will consider different multi-level graphs that are all based on one single graph. This original graph is the train graph TG_{DB} based on the winter 1996/97 train timetables of the German railroad company Die Bahn (DB). It consists of 6960 stations, 931746 vertices, and 1397619 edges.

The second input to the multi-level graph for train graphs is the sequence of sets of stations $\Sigma_1, \ldots, \Sigma_l$, which determines the multi-level graph, referred to by $\mathcal{M}(TG_{DB}; \Sigma_1, \ldots, \Sigma_l)$. In the following we will omit the graph TG_{DB} in the notation of the multi-level graph. The goal of this experimental study is to investigate the behaviour of the multi-level graph with respect to the sequence

$\Sigma_1, \ldots, \Sigma_l$. The experiments to measure the raw CPU time were run on a Sun Enterprise 4000/5000 machine with 1 GB of main memory and four 336 MHz UltraSPARC-II processors (of which only one was used). The preprocessing time to construct a multi-level graph (i.e., the additional edges and the component tree) varies from one minute to several hours.

Parameters. First of all, we want to measure the improvement in performance of shortest path algorithms if we compute the shortest path in the subgraph of $\mathcal{M}$ instead of the original graph G. From the snapshot of over half a million of realistic timetable queries that has been investigated in [20] we take a subset of 100000 queries. Then, for each instance of a multi-level graph $\mathcal{M}$ that we consider we solve the queries by computing the corresponding shortest path in the subgraph of $\mathcal{M}$ using Dial's variant of Dijkstra's algorithm (since this variant turned out to be the most suitable as our previous study [20] exhibited). From these shortest path computations we consider two parameters to evaluate the improvement of the performance:

- *CPU-speedup*: the ratio between the average CPU time needed for answering a single query in the original train graph (0.103 secs) and the average CPU time when the subgraph of $\mathcal{M}$ is used;
- *edge-speedup*: the same ratio when the average number of edges hit by Dijkstra's algorithm is used instead of the average raw CPU time.

Note that the time needed to compute the subgraph for a given query is only included in the CPU-speedup, not in the edge-speedup. Another issue is the space consumption, and therefore we define

- the *size of a level* of $\mathcal{M}$ to be the number of edges that belong to that level;
- the *size* of $\mathcal{M}$ to be the total number of edges in all levels of $\mathcal{M}$ (including the original graph G);
- the *relative size* of $\mathcal{M}$ to be the size of $\mathcal{M}$ divided by the number of edges in the original graph G.

Finally, to compare the improvement in performance and the space consumption, we consider the (CPU-, edge-) *efficiency* of $\mathcal{M}$, being the ratio between (CPU-, edge-) speedup and the relative size of $\mathcal{M}$.

4.1   Two Levels

In the following we define the sequences of sets of stations used in our experiments with 2-level graphs.

We define three sequences $A = (A_1, \ldots, A_{10})$, $B = (B_1, \ldots, B_{10})$, and $C = (C_1, \ldots, C_{10})$ of sets of stations, which are decreasing with respect to set inclusion. The first set in each sequence is identical for all the three and consists of all the stations that have a degree greater than two in the station graph; this yields a set of 1974 stations. The last set of each sequence contains 50 stations, and the sizes of the remaining 8 sets of stations are such that the sizes are equally distributed in the range [50, 1974].

The difference between A, B, and C is the criterion on the selection of stations:

A: In the timetable data, each station is assigned a value that reflects the *importance* of that station with respect to changing trains at that station. The sets A_i contain the stations with the highest importance values.
B: The sets B_i contain stations with the highest degrees in the station graph.
C: The set C_1 is a random set of stations. Then, for C_k ($2 \le k \le 10$), stations are randomly selected from C_{k-1}.

These criteria for selecting stations are crucial for the multi-level graph approach. For criteria A and B we use additional information from the application domain: they reflect properties of important hubs in the railroad network. Removing these hubs yields intuitively a "good" decomposition of the network. The experimental results confirm this intuition.

Using each set of stations A_i, B_i, and C_i ($i = 1, \ldots, 10$) as the set Σ_1, we compute the 2-level graph (i.e., consisting of the original graph being level zero and level one) $\mathcal{M}(\Sigma_1)$. Figure 3 shows the sizes (i.e., the number of edges) of the level one. For the sequences A and B these sizes are similar, and for the randomly selected sets C, the size grows dramatically as the number of stations decrease. In the following we will focus on the sequence A, since B shows similar but slightly worse results, and the multi-level graphs using sets of stations of C are too big.

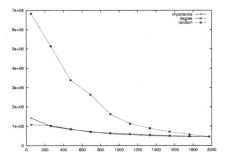

Fig. 3. For each sequence A, B, and C, there is one curve. Each point corresponds to one set Σ_1 of stations in these sequences. The diagram shows the size of level one of the 2-level graph $\mathcal{M}(\Sigma_1)$ according to the number of stations in Σ_1.

For $i = 1, \ldots, 9$ with decreasing number of stations in A_i, the speedup and efficiency of $\mathcal{M}(A_i)$ is growing, and from 9 to 10 it is falling drastically, as Figure 4 shows. Figure 5 reveals one reason for this behaviour: While the number of stations in A_i is big enough, for almost all queries ($> 96\%$) the level one is used, i.e., the subgraph of $\mathcal{M}(A_i)$ used for the shortest path computation consists of the corresponding upward and downward edges of level one, and of the edge-set E_1. But for $i = 10$, for only about 60% of the queries the level one is used, and the remaining 40% of the queries have to be solved in level zero, i.e., on the original graph. The queries for which level one is used still profit from level one as Figure 5 shows, but for the rest of the queries the speedup equals one. In total, this reduces the average speedup over all queries.

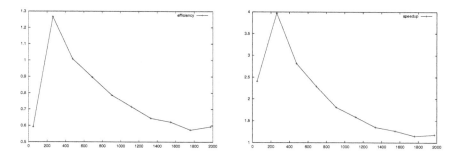

Fig. 4. Each point corresponds to one 2-level graph $\mathcal{M}(A_i)$ for each set of stations in A. The left diagram shows the CPU-efficiency of $\mathcal{M}(A_i)$ according to the number of stations in A_i, and in the right diagram the ordinate is the average CPU-speedup for the 2-level graphs.

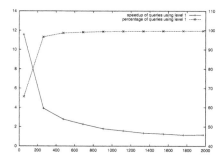

Fig. 5. Like Figure 4, the points refer to sets of stations A_i, and the abscissa denotes the number of stations in A_i. For the curve that is growing with respect to the number of stations, the ordinate on the right shows the percentage of queries *for which the second level is actually used* (i.e., the lowest common ancestor in the component tree is the root), while for the descending curve the average CPU-speedup over all these queries is shown on the left ordinate.

4.2   Multiple Levels

The experiments with two levels show, that the set of stations A_9 with 263 stations yields the best performance, and (according to Figure 5) that the most interesting cases to investigate is to consider subsets with less than $|A_9|$ stations. In our test sequence, there is only the set A_{10} with less stations. Consequently, we included in the sequence A for our investigation with more than two levels also the subsets of stations A_{9a} (225 stations), A_{9b} (156 stations), A_{9c} (100 stations), and A_{10a} (30 stations).

Three Levels. For every pair Σ_1, Σ_2 of sets of stations in A with $\Sigma_1 \supset \Sigma_2$, we consider the 3-level graph $\mathcal{M}(\Sigma_1, \Sigma_2)$. For fixed Σ_1, we investigate the behaviour of the 3-level graph with respect to Σ_2. Figure 6 shows this behaviour for $\Sigma_1 \in \{A_1, A_7, A_8, A_9\}$. With $\Sigma_1 = A_1$, we see the same drop of speedup and efficiency when Σ_2 gets too small as in the 2-level case. However, when the size of Σ_1 decreases (e.g., $\Sigma_1 = A_9$), we observe that the suitable choices for Σ_2 are the subsets A_{10} (50 stations) and A_{10a} (30 stations) which improve both speedup and efficiency. This also shows that different levels require different sizes of subsets.

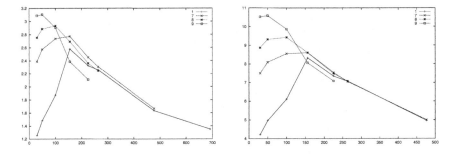

Fig. 6. The equivalent of Figure 4 for 3-level graphs $\mathcal{M}(\Sigma_1, \Sigma_2)$. For each set of stations $\Sigma_1 \in \{A_1, A_7, A_8, A_9\}$ there is one curve, which is obtained by varying Σ_2 (abscissa). On the left hand, the ordinate shows the CPU-efficiency, while on the right hand the CPU-speedup is shown.

More Levels. For more than three levels, we do not investigate every possible combination of sets of stations in A, but follow an iterative approach. To get initial sequences $\Sigma_1, \ldots, \Sigma_{l-1}$ for the l-level graph, we take the sequences $\Sigma_1, \ldots, \Sigma_{l-2}$ that were the basis for the best $l-1$-level graphs, and combine these sequences with the sets of stations Σ_{l-1} in A with $\Sigma_{l-1} \supset \Sigma_{l-2}$. Then, subsequences of A are used as input for the l-level graph that are similar to the initial sequences.

The following table as well as Figure 7 show the results for the best l-level graph for $2 \le l \le 6$. Note that the one-level graph is the original graph, and that the speedup and efficiency are ratios comparing the results for multi-level graphs with the original graph, so for the original graph the speedup and efficiency are one.

Levels	$\mathcal{M}(\cdot)$	speedup		efficiency	
		CPU	edge	CPU	edge
2	A_9	3.97	4.89	1.37	1.56
3	A_9, A_{10}	10.58	14.06	3.11	4.12
4	A_7, A_{9b}, A_{10a}	**11.18**	16.63	**3.48**	5.18
5	$A_7, A_{9b}, A_{9c}, A_{10a}$	9.91	**17.52**	3.06	**5.41**
6	$A_7, A_9, A_{9a}, A_{9c}, A_{10a}$	8.58	17.01	2.55	5.06

The gap between CPU- and edge-speedup reveals the overhead to compute the subgraph $\mathcal{M}_{st}$ for a query using the multi-level graph, since the average CPU-time includes this computation, but the average number of edges hit by Dijkstra's algorithm does not. Considering levels four and five, because of this overhead the CPU-speedup is decreasing while the edge-speedup is still increasing with respect to the number of levels. Experiments with larger values of l revealed that there is no further improvement in the speed-up and/or in the efficiency.

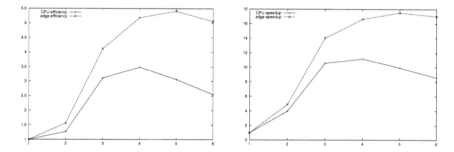

Fig. 7. For different numbers l of levels the results of the best l-level graph is shown: the CPU- and edge-efficiency in the left diagram, and the CPU- and edge-speedup in the right one.

5   Conclusions

In this study, we empirically investigated a hierarchical decomposition approach based on multi-level graphs for a specific application scenario. Given the complexity of the recursive construction of the multi-level graph (or of similar models proposed in the literature), this concept might appear to be more of theoretical interest than of practical use. To our surprise, our experimental study with multi-level graphs for this specific scenario exhibited a considerable improvement in performance regarding the efficient computation of on-line shortest path queries.

In defining the multi-level graphs, we considered three simple criteria A (importance of stations), B (highest degrees), and C (random choice) to select the stations. The latter criterion turned out to be a very bad choice. Further improvements could be possibly achieved by using more sophisticated versions of criteria A and B. For example, in [20], a more sophisticated version of criterion A for 2-level graphs was used. This criterion adds new stations to a set of stations for which the 2-level graph is already known and hence is not applicable to generate sets of stations with fixed sizes for more than two levels. Consequently, it could not directly be used here. However, based on the similarities of the results for the sequences A and B, we believe that if a criterion is chosen which yields a better performance for 2-level graphs, the performance of the multi-level graphs with more than two levels could be improved as well.

It would be interesting to investigate the multi-level graph approach in other contexts as well.

Acknowledgment. We would like to thank Martin Holzer for his help in part of the implementations and experiments.

References

1. R. Agrawal and H. Jagadish. Algorithms for Searching Massive Graphs. *IEEE Transact. Knowledge and Data Eng.*, Vol. 6, 225–238, 1994.
2. C. Barrett, R. Jacob, and M. Marathe. Formal-Language-Constrained Path Problems. *SIAM Journal on Computing*, Vol. 30, No. 3, 809–837, 2000.
3. U. Brandes, F. Schulz, D. Wagner, and T. Willhalm. Travel Planning with Self-Made Maps. *Proc. 3rd Workshop Algorithm Engineering and Experiments* (ALENEX '01), Springer LNCS Vol. 2153, 132–144, 2001.
4. A. Car and A. Frank. Modelling a Hierarchy of Space Applied to Large Road Networks. *Proc. Int. Worksh. Adv. Research Geogr. Inform. Syst.* (IGIS '94), 15–24, 1994.
5. S. Chaudhuri and C. Zaroliagis. Shortest Paths in Digraphs of Small Treewidth. Part II: Optimal Parallel Algorithms. *Theoretical Computer Science*, Vol. 203, No. 2, 205–223, 1998.
6. S. Chaudhuri and C. Zaroliagis. Shortest Paths in Digraphs of Small Treewidth. Part I: Sequential Algorithms. *Algorithmica*, Vol. 27, No. 3, 212–226, Special Issue on Treewidth, 2000.
7. G. Frederickson. Planar graph decomposition and all pairs shortest paths. *Journal of the ACM*, Vol. 38, Issue 1, 162–204, 1991.
8. G. Frederickson. Using Cellular Graph: Embeddings in Solving All Pairs Shortest Path Problems. *Journal of Algorithms*, Vol. 19, 45–85, 1995.
9. http://bahn.hafas.de. Hafas is a trademark of Hacon Ingenieurgesellschaft mbH, Hannover, Germany.
10. M. Müller-Hannemann and K. Weihe. Pareto Shortest Paths is Often Feasible in Practice. *Proc. 5th Workshop on Algorithm Engineering* (WAE'01), Springer LNCS 2141, 185–197, 2001.
11. K. Ishikawa, M. Ogawa , S. Azume, and T. Ito. Map Navigation Software of the Electro Multivision of the '91 Toyota Soarer. *IEEE Int. Conf. Vehicle Navig. Inform. Syst.*, 463–473, 1991.
12. R. Jakob, M. Marathe, and K. Nagel. A Computational Study of Routing Algorithms for Realistic Transportation Networks. *The ACM Journal of Experimental Algorithmics*, Vol. 4, Article 6, 1999.

13. S. Jung and S. Pramanik. HiTi Graph Model of Topographical Road Maps in Navigation Systems. *Proc. 12th IEEE Int. Conf. Data Eng.*, 76–84, 1996.
14. D. Kavvadias, G. Pantziou, P. Spirakis, and C. Zaroliagis. Hammock-on-Ears Decomposition: A Technique for the Efficient Parallel Solution of Shortest Paths and Other Problems. *Theoretical Computer Science*, Vol. 168, No. 1, 121–154, 1996.
15. T. Preuss and J.-H. Syrbe. An Integrated Traffic Information System. *Proc. 6th Int. Conf. Appl. Computer Networking in Architecture, Construction, Design, Civil Eng., and Urban Planning* (europIA '97), 1997.
16. S. Shekhar, A. Fetterer, and G. Goyal. Materialization trade–offs in hierarchical shortest path algorithms. *Proc. Int. Symp. Large Spatial Databases*, Springer LNCS 1262, 94–111, 1997.
17. S. Shekhar, A. Kohli, and M. Coyle. Path Computation Algorithms for Advanced Traveler Information System (ATIS). *Proc. 9th IEEE Int. Conf. Data Eng.*, 31–39, 1993.
18. J. Shapiro, J. Waxman, and D. Nir. Level Graphs and Approximate Shortest Path Algorithms. *Network* 22, 691–717, 1992.
19. L. Siklóssy and E. Tulp. The Space Reduction Method: A method to reduce the size of search spaces. *Information Processing Letters*, 38(4), 187–192, 1991.
20. F. Schulz, D. Wagner, and K. Weihe. Dijkstra's Algorithm On-Line: An Empirical Study from Public Railroad Study. *ACM Journal of Experimental Algorithmics*, Vol. 5, Article 12, 2000, Special issue on WAE'99.
21. J. D. Ullman and M. Yannakakis, High Probability Parallel Transitive Closure Algorithms, *SIAM J. on Computing* 20(1), pp. 100-125, 1991.

Evaluating the Local Ratio Algorithm for Dynamic Storage Allocation*

Kirk Pruhs[1] and Eric Wiewiora[2]

[1] Dept. of Computer Science
University of Pittsburgh
Pittsburgh, PA. 15260 USA
kirk@cs.pitt.edu
http://www.cs.pitt.edu/~kirk
[2] Computer Science and Engineering Department
University of California – San Diego
cwiewior@cs.udsd.edu

Abstract. We empirically compare the local ratio algorithm for the profit maximization version of the dynamic storage allocation problem against various greedy algorithms. Our main conclusion is that, at least on our input distributions, the local ratio algorithms performed worse on average than the more naive greedy algorithms.

1 Introduction

The paper [1] presents a general framework, the local ratio technique, for solving many scheduling and resource allocation problems. The input for these problems consists of tasks/tasks that must be scheduled. These tasks have some resource requirement, and some profit that is obtained if they are scheduled. The most obvious naive algorithms only consider one parameter of the tasks, and thus these naive greedy algorithms can then be arbitrarily bad in the worst case. For example, one possible greedy algorithm would be to schedule the most profitable tasks first, which would be bad if the slightly more profitable tasks have much higher resource requirements. For many of these problems it was not at all clear how to construct simple reasonable algorithms. The local ratio technique can be viewed as "sophisticated" greedy algorithm that balances these competing demands. For some of these problems, the local ratio algorithm is the first known constant approximation algorithm. So in summary, for these problems the local ratio algorithm is generally vastly superior, in the worst case, to the more naive greedy algorithms.

Our goal in this paper is to empirically test whether the local ratio algorithms are also superior on average to the more naive greedy algorithms. The application from [1] that we have selected for our experiments is the dynamic storage allocation problem with the objective function of maximizing the profit of the

---

* Supported in part by NSF grants CCR-9734927, CCR-0098752, ANIR-0123705, and by a grant from the US Air Force.

D. Mount and C. Stein (Eds.): ALENEX 2002, LNCS 2409, pp. 60–70, 2002.
© Springer-Verlag Berlin Heidelberg 2002

scheduled tasks. This problem seemed to us to be one of the more applicable and harder problems in [1].

We implemented several different variations of the local ratio algorithm, and many different greedy algorithms. We implemented a random instance generator, and a graphical user interface to view the outputs produced by the algorithms. All of these programs can be found at:

http://www.cs.pitt.edu/~kirk/dynstore.

An example of the graphical user interface showing a particular schedule is shown in figure 1. The graphical user interface proved to be quite a useful tool. The performance of the various algorithms is quite obvious from a visual inspection of their schedules for a few inputs.

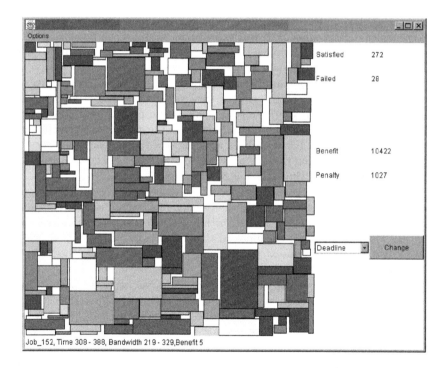

Fig. 1. An example of a flight schedule using a greedy placement

Our main conclusion is that, at least on our input distributions, the local ratio algorithms performed worse on average than the more naive greedy algorithms.

In section 2 we survey the related background, including a formal definition of the dynamic storage allocation problem, and descriptions of the algorithms that we implemented. In section 3 we describe the input distributions that we used. In section 4 we give our experimental results.

2   Background

2.1   Profit Maximization Dynamic Storage Allocation

An instance consists of n tasks, where the ith task R_i has a bandwidth requirement b_i, and a length/duration p_i. Thus each task can be thought of as a rectangle, with vertical height equal to the bandwidth, and horizontal height equal to the length. The goal is to assign the tasks to a scheduling space, which is a larger rectangular space with height normalized to 1, and length equal to some value T. No two tasks can be placed in such a way that they overlap in this scheduling space. Further each task R_i has a release time r_i, and a deadline d_i. Each task R_i must be started after r_i, and completed before time d_i. It is also convenient to define the laxity of a task R_i to be $\ell_i = d_i - r_i - p_i$. Finally, each task R_i has an associated positive integer benefit w_i that is obtained by the scheduler if R_i is scheduled. In the *unit case setting*, each benefit is 1. The objective function is to maximize the aggregate benefit of the scheduled tasks. We assume that all numbers are integers.

Dynamic storage allocation is classic combinatorial optimization problem with a wide variety of applications. For surveys see [2,6]. Essentially all interesting versions of the the dynamic storage allocation problem are NP-hard [2]. The most commonly studied objective function is the size of the spectrum used to schedule all requests. That assumes that the scheduling space that does not have an a priori bound on the size of the spectrum. The best competitive ratio known for a polynomial time offline algorithm is 3 [3].

In [1] the local ratio algorithm is shown to have a constant worst-case approximation ratio for the profit maximization version of dynamic storage allocation. In [5] a LP-based polynomial time algorithm with a constant competitive ratio is given. The authors of [5] encountered this problem in a EU research project Euromednet within the context of scheduling requests for remote medical consulting on a shared satellite channel.

There have been some experimental studies of some heuristics such as First Fit for the objective function of minimizing the bandwidth required to schedule all tasks (see [6] for a survey), but to the best of our knowledge there has been no experimental studies of algorithms for the profit maximization version of dynamic storage allocation.

2.2   The Greedy Algorithms

All of our greedy algorithm consider the tasks in some order. All of the greedy algorithm use First Fit to place the task under consideration, that is, the greedy algorithms differ only in the order that they attempt to place the tasks. First Fit tries to place the task as early in time as possible, and then as low in the bandwidth dimension as possible. We considered about 12 different ordering methods. We only report on the best 2 here.

Deadline : Considers the tasks in nondecreasing order of $r_i + \ell_i$, the latest time
that the task can be run. (Note that this is a nonstandard use of the term
deadline.)

ProfitArea : Considers the tasks in nonincreasing order of $\frac{w_i}{p_i \cdot b_i}$, the benefit
density of the request.

2.3   The Local Ratio Algorithms

We now explain how the local ratio algorithm works on an arbitrary instance.
First $\ell_i + 1$ copies of each task R_i are created in the instance, one for each possible
starting time of R_i. The release time of the jth copy, $0 \le j \le \ell_i$, is $r_i + j$, and
the deadline of the jth copy is $r_i + j + p_i$. So each copy has zero laxity.

The local ratio algorithm consists of two stages. The first stage is recursive
and determines the placement S of the tasks to be scheduled in time. The first
stage has a parameter α, which in some sense captures the value of tasks that
can be finished quickly versus the value of other characteristics of a task. If the
linear resource was fungible (in this case that means that the bandwidth used
by each task could be reassigned at each time instant) then S would be feasible.
However, in our dynamic storage allocation problem, the resource is not fungible.
So a second stage is required to assign particular bandwidths to the tasks.

We now summarize the first stage. First all tasks with non-positive benefits
are removed. Note that while all tasks initially have positive benefits, the benefits
may become zero/negative in the course of the algorithm. If no tasks remain,
then the empty schedule is returned. Among those tasks that can be added to
the current schedule, select a task R_i that can be finished as early as possible.
That is, if one can add R_i to the current schedule so that it is finished at time
t, then the ending time of every other task that one might add to the schedule
would be no earlier than time t. In the key step, the benefits of each task R_j are
modified in the following manner:

- If R_j is a copy of R_i, then w_j is set to $w_j - w_i$. While initially $w_i = w_j$ if
  these are two copies of the same task, the values may diverge in the course
  of the algorithm.
- If R_j overlaps R_i and R_i and R_j are not copies of the same task, then w_j is
  set to $w_j - \alpha \cdot b_j \cdot w_i$. So the more valuable the selected job R_i is, the less
  valuable the overlapping job R_j is. The higher the bandwidth of R_j the more
  its weight is lowered. And α balances these two factors versus the factor that
  R_i may be completed before R_j.
- Otherwise, w_j is not modified.

The problem is solved recursively on the modified instance. The task R_i is added
to the returned schedule if there is sufficient bandwidth to add R_i, and no other
copy of R_i has already been added to the schedule. Otherwise, the returned
schedule is unmodified.

In [1], it is proposed that Kierstead's algorithm [4], for the standard dynamic
storage allocation problem, be used for the second stage. Kierstead's algorithm

schedules all the tasks in S using bandwidth at most 6. The final schedule is then the horizontal slice, of height 1, that maximizes the weight of the tasks fully contained within the slice. We call the resulting algorithm *Unified+K*, as the local ratio algorithm is called the unified algorithm in [1].

We also tried various other possibilities for the second stage. Kierstead's algorithm involves rounding the bandwidths up to the next highest power of two. After these tasks are assigned bandwidth in the second phase, this may leave obvious gaps. In *Unified+KC* these gaps are removed, by letting the tasks drop in the bandwidth space as if by gravity, before the best horizontal slice is selected.

The algorithm *Unified+FF* considers the tasks from the first stage in random order, and then assigns the tasks to bandwidth using First Fit. Again the final schedule is the horizontal slice of height 1 that maximizes the weight of the tasks fully contained within the slice.

Unified+FF2 is the same as Unified+FF, except that bandwidth requirements are rounded to the next higher power of 2. After the assignments are performed, the requests are returned to their original bandwidth. The resulting extra space is removed before the best bandwidth range is found.

It will also be instructive to plot the value of the tasks selected by the local ratio algorithms in the first stage. We will call this schedule simply *Unified*. Note that Unified is not a feasible schedule.

3   The Input Distribution

In this section we describe how our inputs were created. Recall that T is the time period to be scheduled. Our input generalization is parameterized by 4 variables:

- n: The number of tasks
- C: The minimum bandwidth required to schedule all of the tasks
- L: The maximum ratio of a task's laxity to its length.
- W: The range of benefits of the tasks

We began with one task that has length T and bandwidth C. We then iteratively cut this large task into smaller tasks until the desired number n of inputs was achieved. We will think of the T by C task as being in the bottom of the scheduling space, and this placement will be an optimal solution if the total bandwidth measure $C \leq 1$. Generating instances with known optimum solutions is a standard method that allows one to compare algorithms against the optimal solution, as well as against each other.

On each iteration, all of the tasks from the previous round are split into two tasks. On the first round, and subsequent odd rounds, the tasks are cut horizontally. That is, the aggregate bandwidth of the two resulting tasks are equal to the bandwidth of the original task. On even rounds, the cut is vertical, reducing the duration of the tasks. At the end of every round, the task order is randomized. This process will continue until n tasks are created. The point to cut each task is selected uniformly at random from the middle half of the task to be divided. Thus each partition has at least $1/4$ of the original task. Also, if the input is too small to be divided, that is the task is passed up and the next task is divided.

This sets the length p_i and the bandwidth b_i of each task. If $C \leq 1$, then also sets the placement of the tasks in the optimal schedule. We now explain how to set the release time and deadline for each task. The laxity ℓ_i of each task R_i is selected uniformly at random from the range $[0, L \cdot p_i]$. We thus assume that, given a fixed L, the expected laxity of a task is proportional to its length. The release time r_i of each task is then the start time in the optimal schedule minus a number uniformly at random from the range $[0, \ell_i]$, or zero, whichever is larger. Setting the release date and the laxity fixes the deadline. If the deadline is greater than T, then it is rounded down to T, and the release time is moved back in time so that the laxity remains fixed.

We should note that we also tried input distributions where release times, length and laxity were generated uniformly at random over fixed ranges. This input distributions produced essentially the same results as for the input distribution that we report on here.

4   Experimental Results

We only report on the quality of the solutions produced by the algorithms because all algorithms can be implemented efficiently. The bottleneck in the local ratio algorithms is the first stage. And despite some reasonable effort to implement the first stage efficiently (taking some suggestions from [1]), the local ratio algorithms were clearly slower than the greedy algorithms. We report on a representative sample of the many experiments that we ran.

4.1   Selecting α

Our first goal was to determine how to select the parameter α in the local ratio algorithms. We ran the following tests in order to gauge different α values.

- $n = 300, 400, 500$. These input sizes were sufficiently large for the asymptotic behavior of the algorithms to manifest itself.
- $C = 0.8, .9, 1.0, 1.1, 1.2$. Note that loads near 1 are the most interesting case. If the load is much less than 1, then the problem is easy to solve optimally. In our experiments, inputs with loads below .8 could be solved optimally by at least some of the greedy algorithms. If the load is much more than 1, then, regardless of the scheduling algorithm, the system will reject many tasks, which will probably be unacceptable.
- $L = 0.0, 0.5, 1.0, 1.5, 2.0$
- $W = 1, 40, 80, 120, 160$
- $\alpha = 0.25, 0.50, 0.75, 1.00, 1.25, 1.50, 1.75, 2.00$
  Recall that α is a parameter of the local ratio algorithms.

We generated 10 instances at each point in the experimental space.

The horizontal axis in figure 2 represents the various choices of α. The vertical axis in all of the figures is the average percent of total benefit obtained by the plotted algorithm (not the actual benefit obtained by the algorithm), where the average is taken over all inputs in our experimental space. The vertical bars in all plots are .95 confidence intervals.

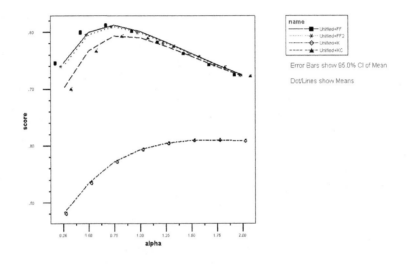

Fig. 2. α versus score

From figure 2 one can see that the best choice for α is about .75 for our sampling space. Thus for the rest of the paper, we report on the local ratio algorithms with α fixed to be .75. We also concluded that Unified+K proposed in [1] is not competitive with the other three local ratio variations, which are all quite similar (at essentially all points in the experimental space). Thus from here on we will only plot Unified+FF.

After having set the best overall value for α, we conducted another round of tests looking into other factors that affect performance in more depth. The rest of the findings of the paper are based on these tests

- $n = 300$
- $C = 0.8, .9, 1.0, 1.1, 1.2$
- $L = 0.0, 0.25, 0.5, 0.75, 1.0, 1.25, 1.5, 1.75, 2.0$
- $W = 1, 20, 80$
- $\alpha = 0.75$

We generated 100 instances at each point in the experimental space.

4.2   The Effect of Weights

The horizontal axis in figure 3 represents values of the weight parameter W, and the vertical axis is the average percent benefit.

Figure 3 illustrates that for all of the algorithms that we tested that the asymptotic effect of the weights kick in for relatively small values of W. So for ease of presentation, from here on, we only present results for the unit weight case $W = 1$ and for $W = 80$, which represents the asymptotic case.

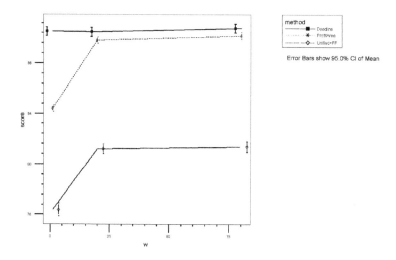

Fig. 3. The Effect of Weight

Since most of the experimental space contains relatively large values for the W parameter, one might think that our choice of $\alpha = .75$ might not be optimal for small W, but in fact this is not the case; $\alpha = .75$ also nearly optimizes the performance of the local ratio algorithms when W is fixed to 1. Fore-shadowing slightly, one can also observe in figure 3 that the local ratio algorithm do not perform as well as the best greedy algorithms.

4.3   The Effect of Laxity

The horizontal axis in figure 4 represents values of the laxity parameter L, and the vertical axis is the average percent benefit. Figure 4 shows illustrates that for all of the algorithms that we tested, laxity didn't have an appreciable effect on the benefit produced. This is somewhat surprising as increasing L increases the flexibility of the scheduler.

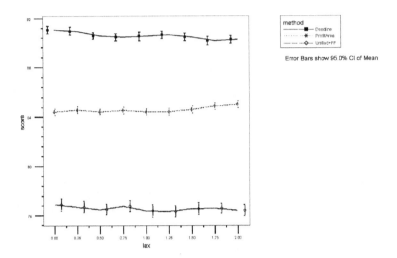

Fig. 4. Laxity L versus Score for $C = 1$ and $W = 1$.

To simplify exposition, we will fix the laxity parameter L to be 1 for the rest of the paper. Note that one can also again see that the greedy algorithms are outperforming the local ratio algorithms.

4.4   Performance Evaluations

We now plot the performance of the two best greedy algorithms, the best unified algorithm, and the value of tasks selected in the first stage of the local ratio algorithm for the best choice of $\alpha = .75$, for a representative choice of $L = 1$, and for $W = 1, 80$.

One can see from figure 5 the best algorithm for $W = 1$ is the Deadline greedy algorithm. In fact, for all but very high loads, the Deadline algorithm accrues even more benefit than the local ratio algorithm accrues in the first stage. Further the greedy algorithms do better than all unified algorithms. Note that for input with local parameter $C < .8$, the Deadline algorithm is quite near optimal.

One can see from figure 6 that, in the case of weights, the Deadline algorithm is the best choice for an underloaded system ($C \leq 1$), and that ProfitArea algorithm is the best choice for an overloaded system ($C > 1$). Although the first stage of the local ratio algorithm beats the greedy algorithms when the load is high (say $C > 1.05$). Recall though that the first stage of the local ratio algorithm does not produce a feasible schedule.

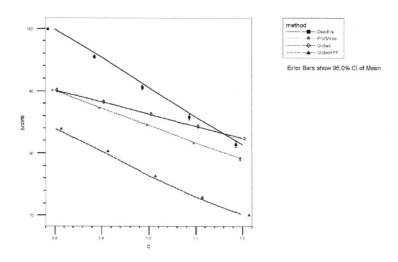

Fig. 5. Load C versus Score for $L = 1$ and $W = 1$

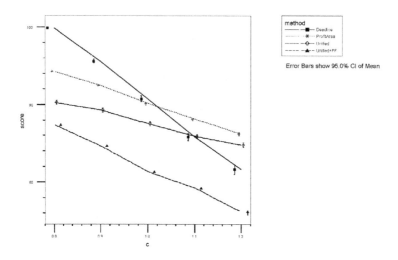

Fig. 6. Load C versus Score for $L = 1$ and $W = 80$

5 Conclusions

Our goal was to compare the more sophisticated local ratio algorithm to more naive greedy algorithms for the profit maximization version of the dynamic spectrum allocation problem. While the local ratio algorithms are better in the worse case, our results suggest that they are inferior on the average case. Further, the local ratio algorithm is harder to program, and and is not as efficient as the greedy algorithms. Our results would suggest against using the local ratio algorithm in practice. It seems that the main reason that the local ratio algorithms does not do as well as the greedy algorithms is that the local ratio algorithms occasionally pick two copies of the same interval in the first stage. While the local ratio algorithms may perform better on highly overloaded systems, but it is not clear how useful that is in practice.

Acknowledgments. This research was conducted while the second author was a student at the University of Pittsburgh.

References

1. A. Barnoy, R. Bar-Yehuda, A. Freund, J. Naor, and B. Schieber, "A unified approach to approximating resource allocation and scheduling", ACM Symposium on Theory of Computing, 2000. To appear in *JACM*.
2. E. Coffman, "An introduction to combinatorial models of dynamic storage allocation", *SIAM Review*, **25**, 311 – 325, 1999.
3. J. Gergov, "Algorithms for compile-time memory allocation", ACM/SIAM Symposium on Discrete Algorithms, S907-S908, 1999.
4. H. Kierstead, "A polynomial time approximation algorithm for dynamic storage allocation", *Discrete Mathematics*, **88**, 231 – 237, 1991.
5. S. Leonardi, A. Marchetti-Spaccamela and A. Vitaletti, "Approximation algorithms for bandwidth and storage allocation problems under real time constraints", Conference on Foundations of Software Technology and Theoretical Computer Science (FSTTCS), 2000.
6. P. Wilson, M. Johnstone, M. Neely, and D. Boles, "Dynamic storage allocation: a survey and critical review", International Workshop on Memory Management, *Lecture Notes in Computer Science*, **986**, 1 – 116, 1995.

An Experimental Study of Prefetching and Caching Algorithms for the World Wide Web*

Massimiliano Curcio, Stefano Leonardi, and Andrea Vitaletti**

Dipartimento di Informatica Sistemistica, Università di Roma "La Sapienza", via Salaria 113, 00198-Roma, Italia.
{leon,vitale}@dis.uniroma1.it

Abstract. Caching and prefetching have often been studied as separate tools for enhancing the access to the World Wide Web. The goal of this work is to propose integrated Caching and Prefetching Algorithms for improving the performances of web navigation. We propose a new prefetching algorithm that uses a limited form of user cooperation to establish which documents to prefetch in the local cache at the client side. We show that our prefetching technique is highly beneficial only if integrated with a suitable caching algorithm. We consider two caching algorithms, Greedy-Dual-Size [6,17] and Least Recently Used, and demonstrate on trace driven simulation that Greedy-Dual-Size with prefetching outperforms both LRU with prefetching and a set of other popular caching algorithms.

1 Introduction

Document *caching and prefetching* are the most widely applied techniques to enhance the access to the World Wide Web.

Caching web documents aims to store frequently accessed documents in a position closer to the user. A first common form of caching is at HTTP proxy servers. A HTTP request is first sent to the proxy server. If the requested document is served from the proxy cache, this is directly sent to the client, otherwise the proxy retrieves the page from the server and then sends it to the client. A second common form of caching is at the client side. The local cache is managed by a web browser to store the more frequently accessed documents. Serving the requests at the local cache may take 10 milliseconds compared with 500 milliseconds average TCP/IP round trip time necessary to open a connection with a remote server. Caching at the client side is even of higher importance for terminals of low bandwidth and memory capability like cellular phones or palmar mobile terminals.

---

* Partially supported by the IST Programme of the EU under contract number IST-1999-14186 (ALCOM-FT), IST-2000-14084 (APPOL), and by the Italian Research Projects MURST 40% "Algorithms for Large Data Sets: Science and Engineering" and "Resource Allocation in Computer Networks".
** Partially supported by Etnoteam (Italy).

D. Mount and C. Stein (Eds.): ALENEX 2002, LNCS 2409, pp. 71–85, 2002.
© Springer-Verlag Berlin Heidelberg 2002

A caching policy is defined by a page replacement rule that selects, when the cache is full, the document to evict. Caching web documents differs in many aspects from traditional caching. In traditional caching, pages have uniform size and every page fault has a uniform cost. The objective is often to reduce the number of page faults through the exploitation of locality of reference in the request sequence. Differently from traditional caching, web caching deals with a dramatic difference from document to document in size and retrieval cost (that are even not necessarily related). The goal of web caching, is not limited to the reduction of cache misses. Web caching also aims to reduce user perceived latency, that depends from the retrieval cost of the documents that are not cached, and network load, the amount of traffic generated from requests not served from cache. There are also some limitations to caching. Caching applies only to static documents, that is however still a good share of the total. Moreover cached documents must be retrieved again if expired. This may be checked at regular intervals or at the time the document is requested by sending a HEAD command to the web server.

Traditional caching strategies like Least Recently Used (LRU), that do not consider document size or retrieval cost, are also widely used in proxy caching (e.g. Squid[1]). Very natural criteria to employ to design more efficient web caching algorithms are to evict documents of big size or low retrieval cost. Cao and Irani [6], and independently Young [17], proposed a simple and elegant cache replacement algorithm, *Greedy-Dual-Size* (GDS), that tries to evict a page with a combination of low retrieval cost, large document size and reference further in the past. Cao e Irani also make an extensive comparison of web caching algorithms on trace driven simulations and observe that Greedy-Dual-Size behaves better than other page replacement algorithms in terms of hit rate, reduced latency and network load. A theoretical analysis proving the efficiency from a worst case analysis point of view of Greedy-Dual-Size can be found in the work of Young [17] and of Cao and Irani [6]. Off-line approximation algorithms for general web caching problems have been recently proposed by Albers et al. [1] and by Bar-Noy et al. [3].

Document prefetching is a second effective tool for improving the access to the World Wide Web. Prefetching aims to bring documents at the client side before they are requested. Prefetching can be initiated either at the client side or at the server side (in this last case prefetching is often denoted by pushing). A key ingredient of a prefetching technique is a document selection policy that rules the choice of which document to prefetch. Different prefetching selection policies have been proposed in literature. For instance in [10] some (or all) the links in the document requested by the user are also sent to the client. In [7] and [14] the clients gather information from the server on the most popular links to decide which documents to prefetch, while Bestavros [4] implemented a system in which the server responds to a request with both the requested page and a list of related resources. Mogul and Padmanabhan describe a similar method of prefetching using server suggestions [15] in which the server predicts the next

---

[1] http://www.squid-cache.org

access on the basis of the weight of a dependency graph, while the client decides whether or not to actually prefetch the page. In [16]the authors implemented a deterministic static client initiated prefetching policy, useful to prefetch daily used pages like news or meteo forecasts. Fan, Jacobson, Cao and Lin [8] consider server initiated prefetching between proxy and low bandwidth modem users. The prefetching technique uses a page predictor at the proxy cache that decides which pages to prefetch on the basis of the analysis of the most recently requested pages. The authors show how prefetching can lead to a saving of up to about 14.6% of the latency, and of 28,6 % if combined with Delta encoding. They assume the LRU caching replacement strategy at the local cache. In their model prefetching may happen even before the user has requested a document for the first time.

Less work has been done on the integration of caching and prefetching techniques. Kroeger, Long and Mogul [12] study the combined effect of caching and prefetching on end user latency. They conclude that caching alone may save user latency up to 26%, while the combined effect of caching and prefetching can lead up to 60% latency reduction. These figures must be intended as a tentative to establish an upper bound on the performance improvement of prefetching and caching in a ideal framework where memory cache is unlimited and full knowledge about the documents that will be requested is assumed. In [9] the authors investigate the use of the dependencies between resources in making cache replacement decisions. In particular combining knowledge of client behaviour with resource dependencies can be useful to reduce latency by prefetching selected resources.

Differently from previous works, we concentrate on the combined and integrated application of prefetching and caching strategies. Integrated prefetching and caching has recently received a growing attention in the attempt of enhancing the performances of local data base and file systems. Cao, Felten, Karlin and Li [5], proposed prefetching/caching strategies to optimize the access to memory blocks stored on magnetic disks. The model considers equal block size and block fetching time from disk to cache. They propose and analyze, both theoretically and empirically, different strategies to decide the time at which the prefetching of the next referenced block not in cache must be initiated. In a later work Albers, Garg and Leonardi present for the model introduced in [5] a first polynomial time prefetching/caching algorithm that achieves the optimum elapsed time. Such strategies have also been extended to parallel disk systems, where more blocks can simultaneously be fetched [2] [11] from disks to cache.

Our Contribution. In this work we propose a new technique to integrate prefetching and caching for the access to web documents. Our attention is focused at the client side. The core of our proposal is the definition of a document selection policy to determine which documents to prefetch and a technique of integration between prefetching and caching. Although our prefetching technique can be integrated with any caching policy, we will concentrate on two specific caching algorithms, Least Recently Used and Greedy-Dual-Size, the first being the most diffused caching strategy, the second being recently shown to be very effective for caching web pages [6].

We compare our proposed prefetching/caching techniques on trace driven simulation with several popular caching algorithms. The conclusion we draw is that the integration between prefetching and caching is very effective only if a suitable caching algorithm is adopted. In particular we demonstrate that the integration with Greedy-Dual-Size outperforms the integration with LRU and all the caching algorithms we consider.

The document selection policy we propose does not involve any complex prediction mechanism. It relies on a simple form of cooperation between user and system to gather a partial knowledge of the documents that will be requested in the future. The user is simply asked to *mark* a page when abandoned for a newly requested page if it intends to request the page again in the future. Knowing that a page will be requested is a valuable information. However, this does not say when the page will actually be requested, i.e. when the page can conveniently be prefetched. For this purpose we exploit locality of reference. All marked pages that are not in cache, i.e. evicted after the last reference, are organized in a priority queue ordered by the time the pages have been referenced last. *The next page to be prefetched is then the most recently used marked page that is not in cache.*

The intuition behind the goodness of this strategy is that this is an optimal choice for selecting the page to prefetch if the user follows a Depth First Search of a hyperlinked structure, i.e. the user follows a hyperlink or uses the back button. Of course the form of cooperation we expect from the user is an ideal one, we assume that, in his interest, he will always say the truth. However we believe that even with only a partial users cooperation the system would improve its performances. We also consider that only a fraction of the documents are cacheable and that expired documents must be retrieved again from the server.

We incorporate our integrated prefetching/caching policy into the web caching simulator developed by Pei Cao and Gideon Glass and used by Cao and Irani [6] for comparing web caching algorithms. We then perform trace driven comparisons between the prefetching/caching algorithms we propose and the web caching algorithms implemented in the simulator. The comparison is done on WWW traces from Berkeley Home IP users used by Cao et al. [8] for experimenting proxy initiated prefetching techniques[2], and traces collected daily within the NLANR project[3].

Structure of the Paper. In Section 2 we present our integrated prefetching/caching algorithm. In Section 3 we describe the web caching simulator and the caching algorithms we compare with. In Section 4 the traces used in the simulation are presented. Section 5 describes the results of the simulation. The whole set of diagrams describing the results of the simulation are in Appendix .

---

[2] http://www.acm.org/sigcomm/ita/
[3] http://www.nlanr.net

```
Let A be a caching algorithm;
Let p be the requested page;
Let init(p, A) be the inizialization phase for page p when cached by A;
Let PQ be the Priority Queue of non-cached marked pages,
ordered by time of last reference to the page. ;
REMOVE(PQ) returns the page referenced last in PQ;


if ((p is in cache) and (p is not expired)) retrieve p from cache;
else retrieve p from server;

// The user abandons page p and it is asked to mark it.

if (p is in cache) and (p is marked by the user) {
   p remains in cache;
   init(p, A);
}

if (p is not in cache) and (p is marked) {
   if (p is cacheble) {
      insert p in cache;
      init(p, A);
      while (the overall size of cached documents exceeds cache memory size) {
            select a page q to evict applying A's eviction policy;
            INSERT(q, PQ);
      }
   }
   else discard p;
}

if (p is not marked) discard p;

if ((PQ is not empty) and (r = REMOVE(PQ) fits the empty space in cache)) {
   prefetch r;
   insert r in cache;
   init(r, A);
}
```

Fig. 1. The prefetching/caching algorithm.

2   Integrating Prefetching and Caching

Our document selection policy is based on a form of cooperation between the user and the system. Every time a document is abandoned the user is asked to mark the document if it intends to request it again in the future. In that case the document is said *marked*, otherwise it is said *unmarked*. The integration of the prefetching technique with any caching algorithm is done as follows:

1. A requested document is retrieved from the server if either not cached or if cached but expired.
2. A cacheable document is cached if marked by the user.
3. A document retrieved from cache or from the server is discarded if not marked.

4. The eviction rule is invoked when the overall size of cached documents exceeds the size of the cache. Observe that even the document cached last can in principle be evicted.
5. The document that is prefetched is the most recently used non-cached document that is marked. Observe that all non-cached marked documents have been previously evicted from cache.

In Figure 1 we give a formal description of the algorithm.

The set of documents that are marked but not cached are organized in a priority queue ordered by time of last reference to the document. The document referenced most recently in the past is then selected for prefetching.

We consider two caching algorithms, LRU and Greedy-Dual-Size. *We denote by GDSP the integration of prefetching with Greedy-Dual-Size and by LRUP the integration of LRU with prefetching.*

The page replacement rule of LRU consists of evicting the page accessed furthest in the past. For this purpose, LRU orders all cached pages in a list. The initialization phase of LRU consists in bringing a cached document at the top of the list. The eviction policy of LRU selects the document at the rear of the list. Observe that if we integrate the prefetching policy with LRU, the priority queue use by the document selection policy can be simply replaced by a stack.

Greedy-Dual-Size [6] [17] considers the retrieval cost, the size of the document and temporal locality in the eviction rule. Let c_p be the retrieval cost of document p and let s_p be the size of document p. A value H_p is associated with every page p in cache. When a page is cached, or it is retrieved from cache, H_p is set to c_p/s_p. When the cache is full, a page with $H_p = 0$ is selected for eviction, if any. Otherwise we compute the minimum ratio $r = min_p H_p$ and reduce every H_p by r. At least one page p will have $H_p = 0$ after this operation is performed. Figure 2 describes the initialization phase and the eviction phase of Greedy-Dual-Size.

```
Set L = 0 at the beginning of the algorithm;

init(p, GREEDY − DUAL − SIZE) {
    Let H_p = L + c_p/s_p;
}

// returns the page to be evicted
page evict − policy(GREEDY − DUAL − SIZE) {
    Let L = min_{q∈cache} H_q;
    return q such that H_q = L;
}
```

Fig. 2. Greedy-Dual-Size

3   The Web Caching Simulator

We integrate Greedy-Dual-Size with Prefetching and LRU with Prefetching within the web cache simulator designed by Pei Cao and Gideon Glass[4] used to compare on HTTP traces several caching strategies. We therefore simulate the marking of the pages operated by the user by traversing the trace and observing which pages are requested in the reminder of the sequence. The data structures used in the simulator are also modified to include the information required for our prefetching/caching algorithms to work[5].

All distinct web objects requested by a user within a trace are stored in a vector of records. The record stores the page ID of the object, its size, its retrieval cost, its "last modified" time, its "user's mark" attribute value, its "cacheable" attribute value, the number of requests for this object so far and the UNIX timestamp of the current request.

We compare our prefetching/caching algorithms with several of the most effective web caching strategies, in particular those experimented in [6] and implemented in the simulator. We consider the following web caching algorithms, in addition to Greedy-Dual-Size and LRU already described in the previous section:

LRU Threshold. This algorithm is a modified version of LRU in which documents larger than a given threshold are never cached.

Size. The eviction rule selects the cached document of biggest size. The objective is to avoid the replacing of a large number of small documents with a document of large size.

Hybrid: The idea of *Hybrid* is to incorporate an estimation of the current network conditions into the caching algorithm. For each document i, the algorithm calculates a function of the connection time to the server storing i, $clat_{ser(i)}$, the bandwidth of the network link $cbw_{ser(i)}$, the document size s_i and the number of references to document i since the last time it was cached $nref_i$. The function is

$$(clat_{ser(i)} + W_B/cbw_{ser(i)}(nref_i^{W_n}))/s_i,$$

where W_B and W_N are constants. The document that is evicted is the document in cache that minimizes the above function.

Least Relative Value (LRV). The value of a document is a function of the probability P_r that the document is accessed again. The *LRV* algorithm selects the document with Lowest Relative Value as the most suitable candidate for replacement. In [13] it is shown how P_r can be efficiently calculated on-line.

In the implementation of LRU with Prefetching, the cache is simulated using a double linked list where each entry is a record type variable associated with a page. Two variables point at any time at the least recently requested page

---

[4] The source code of the simulator (in its original form) is available at :
ftp://ftp.cs.wisc.edu/pub/cao/webcache- simulator.tar.Z
[5] The new web cache simulator is available at the following URL:
http://www.dis.uniroma1.it/curcio/simulator/

(LRU page) and at the most recently requested page, respectively. The marked pages evicted from cache are simply organized in a stack rather than in a priority queue. In fact, a page that is evicted from cache is always inserted at the head of the priority queue. The stack is also implemented with a linked list.

In the implementation of Greedy-Dual-Size with Prefetching, the cache is simulated using a priority queue ordered by the value of H_p as defined by Greedy-Dual-Size. The set of marked pages that are evicted are also organized in a priority queue ordered by the time of last reference to the page. Both priority queues are implemented with linked lists.

4   HTTP Traces

As the results from a trace-driven study strongly depend on the traces, we gathered two meaningful set of web traces:

- Berkeley University HomeIP proxy traces[6] (1- 19 Nov 1996), containing four sets of traces, for a total of 5,816,666 registered access from 7,714 users (after pre-processing).
- NLANR daily proxy traces from IRCache group[7]: we considered traces from 1 day of HTTP activity (9/20/2000), for a total of 981,354 registered access from 142 users (after pre-processing).

Berkeley traces show an average number of about 7800 requests per user trace, of which, on average, 16% is non-cacheable. The average number of requests that are referenced only once is about 3600, the so called *one timers*. Berkeley traces are also characterized by a quite large number of documents of big latency due to the slow speed of the modems used from users to connect, from 4 kb to 24 kb.

NLANR traces present a number of about 17720 requests per user, with 15000 distinct requests, 14300 one timers and a percentage of about 31% of non-cacheable requests. These traces contain a much larger number of one timers with respect to Berkeley traces and therefore we expect that the benefit of prefetching and caching will be limited in absolute terms.

We performed some necessary preprocessing over the traces. We considered only HTTP requests, and we left out all the requests with essential informations (such as timestamp) missing and all requests with HTTP commands different from GET and POST.

We sorted traces by timestamp and pasted together all files containing a single user's browsing session, obtaining a single file for each user. This doesn't "misrepresent" user's behaviour, because we assume the use of persistent caching (files remain in cache between consecutive browsing sessions, unless user explicitly deletes them).

---

[6] http://www.acm.org/sigcomm/ita/
[7] http://www.nlanr.net

All requests for URLs whose pathname contains the string "cgi-bi" or the character '?', meaning that server's response is a dynamic page or the result of a query submitted by the user, have been considered non-cacheable by setting to 0 the cacheable field, while in all other requests we set this field to 1. We also set the cacheable field to 0 for all the requests with HTTP status different from 200, 203, 206, 300, 301. With a simple modification to the simulator's source code we made it to control this field's value before deciding to insert or not the page currently requested by the user.

A critical issue is the simulation of user's decisions to mark or not to mark the requested pages. The user's decision is simulated by setting to 1 or to 0 a dedicated field of the record corresponding to that request. A simple but quite effective way to simulate this decisional mechanism is to use our "foreknowledge" of the whole sequence of user's requests, obtained by parsing the traces file, to find the web objects that have been requested more than once by the user after the first access.

5 Simulation Results

In this section we present the results of the simulation of the prefetching and caching algorithms on the set of HTTP traces described in the previous section. We measure the performances of our algorithms in terms of the three following metrics:

- Hit rate: the number of requests that hit in cache as a percentage of the total number of requests.
- Byte hit rate: the number of bytes that hit in cache as a percentage of the total number of bytes requested.
- Reduced latency: the percentage of total retrieval time saved by the algorithm, i.e. the ratio between the overall retrieval time (from servers) of documents served from cache and the overall retrieval time of documents retrieved from servers.

Algorithms GDSP and LRUP are evaluated on these three metrics together with the caching algorithms described in Section 3, namely LRU, $LRU - Threshold$, $Size$, $Hybrid$, GDS and LRV.

We also partition the set of traces by their length. Intuitively, longer sequences should allow to gather more information about the documents requested in the future by the user. For this purpose, we distinguish between traces with a number of requests in the range 5,000-10,000, 10,000-25,0000 and 25,000-50,000.

In appendix we show the results of the simulation. In the following we give a brief description of these results mostly referring to Berkeley Traces. Similar observations can be done for NLANR traces even though the figures are quite different given the different nature of the traces.

As a benchmark for our comparison we consider an *optimal algorithm* having cache size sufficient large to store all the distinct documents in a trace. The optimal algorithm is charged with the retrieval cost of a document only at (1.)

the first reference of a document; (2.) any later reference of a document if expired (3.) any later reference of a document if not cacheable. The simulation has been carried out by monitoring each algorithm on different cache memory sizes, expressed as percentage of the *optimal cache size*. This means that the performance of any algorithm converges to the optimum with the increase of the percentage of the optimal cache size that is available.

The goal of our simulation is therefore to establish the rate of convergence of the various algorithms with the increase of the cache size for each of the three metrics. However, in the diagrams shown in appendix, we restrict our attention, for reasons that will be clear soon, up to 20% of the optimal cache size.

The most remarkable result is that GDSP converges to the optimum hit rate and byte hit rate with about 5% of the optimal cache. GDSP also converges to the optimum reduced latency in most cases with 5% to about 10% of the optimal cache. All other algorithms, including LRUP, converge to the optimum with more than 20% of the optimal cache.

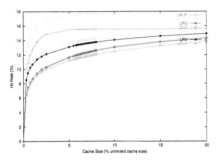 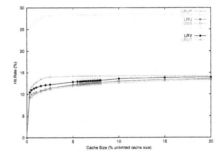

Fig. 3. Berkeley (5000-10000 users): Hit Rate.   **Fig. 4.** Berkeley (10000-25000 users): Hit Rate.

For example, in Figure 8, the value of byte hit rate converges to 22% with 5% of the optimal cache, while all other algorithms are still below a byte hit rate of 10% with more than 20% of the optimal cache. A similar behaviour is observed for hit-rate and reduced latency. For example in Figure 4 GDSP converges to a hit rate of 27% with 3% of the optimal cache, in Figure 10, to a reduced latency of about 14% with 10% of the optimal cache. We stress that the values of convergence are very close to the optimum, that is with 100% of the optimal cache size.

It might appear very surprising that GDSP achieves the optimum with only 5% of the optimal cache. This is explained with the large number of non-cacheable documents and one-timers in the traces. In Berkeley traces these two kind of documents are about 70% of the total number of requests, even higher figures are reached for NLANR traces. The reason for this behaviour is that the optimal cache size also includes the size of non-cacheable documents and one-timers, while GDSP and LRUP only cache those documents requested again in

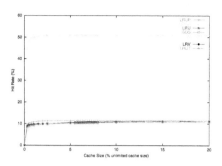

Fig. 5. Berkeley (25000-50000 users): Hit Rate.

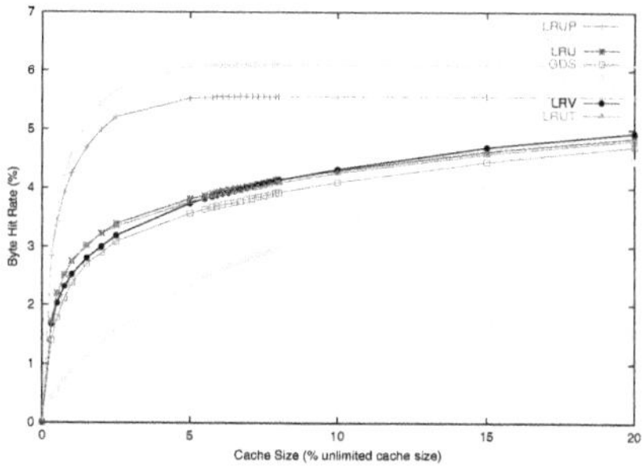

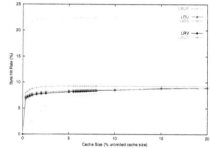

Fig. 6. Berkeley (5000-10000 users): Byte Hit Rate.

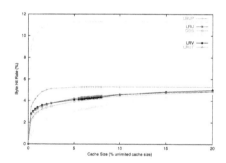

Fig. 7. Berkeley (10000-25000 users): Byte Hit Rate.

Fig. 8. Berkeley (25000-50000 users): Byte Hit Rate.

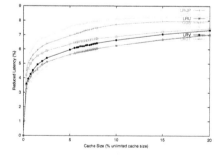

Fig. 9. Berkeley (5000-10000 users): Reduced Latency.

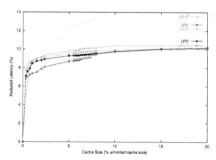

Fig. 10. Berkeley (10000-25000 users): Reduced Latency.

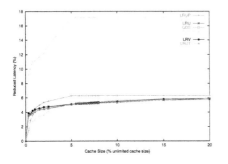

Fig. 11. Berkeley (25000-50000 users): Reduced Latency.

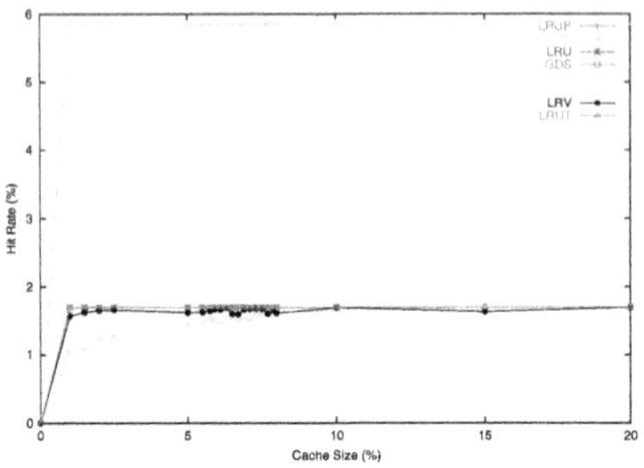

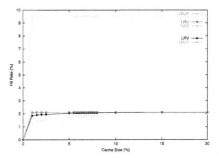

Fig. 12. NLANR (5000-10000 users): Hit Rate.

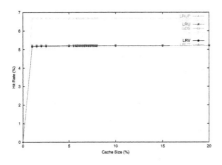

Fig. 13. NLANR (10000-25000 users): Hit Rate.

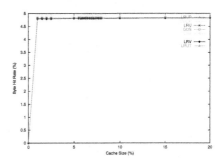

Fig. 14. NLANR (25000-50000 users): Hit Rate.

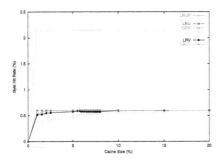

Fig. 15. NLANR (5000-10000 users): Byte Hit Rate.

Fig. 16. NLANR (10000-25000 users): Byte Hit Rate.

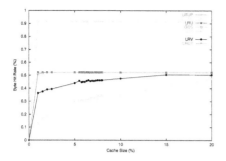

Fig. 17. NLANR (25000-50000 users): Byte Hit Rate.

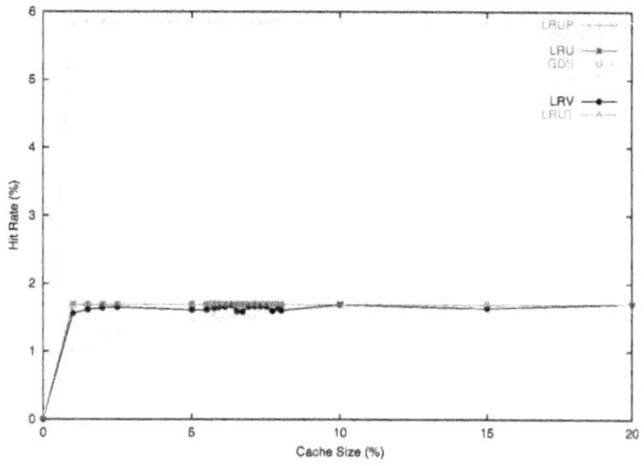

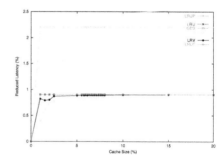

Fig. 18. NLANR (5000-10000 users): Reduced Latency.

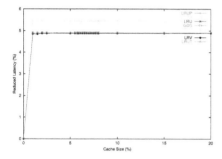

Fig. 19. NLANR (10000-25000 users): Reduced Latency.

Fig. 20. NLANR (25000-50000 users): Reduced Latency.

the reminder of the sequence. This also explains as in many cases LRUP converges faster than all other caching algorithms, even if to a value close to that reached by the other caching algorithms in the interval up to 20% of the optimal cache.

We would also like to comment on the absolute values of our metric of performances. For instance GDSP converges in Figure 4 to a hit rate of about 28% with about 5% of the optimal cache. This should be interpreted as follows. In that sequence only about 40% of the requests are cacheable. GDSP does not hit the first request to a cacheable document but, as the optimum does, retrieves from cache almost all the later references to cacheable documents. In this specific example, we observe that LRUP and all the other caching algorithms hit at most 15% of the requests with 20% of the optimal cache, therefore only about 50% of the requests that are served from cache from GDSP and the optimal algorithm.

All the considerations made for Berkeley traces can also be applied to NLANR traces. For NLANR traces we observe that the absolute values for the three metrics are much smaller than for Berkeley traces. For example, for sequences with 5,000 - 10,000 requests, GDSP approaches a hit rate of about 6%

with 2% of the optimal cache (see Figure 12), while in Berkeley traces, for sequences of length in the same range, a hit rate of 18% is reached. Of course the explanation is in the bigger number of non-cacheable documents and one-timers with respect to Berkeley traces. The effect of a large number of these two kind of documents is twofold: 1) it decreases the cache size needed to converge to the optimum, since the optimal algorithm caches all the pages; 2) decreases the optimal performances, since only a smaller percentage of the documents can be retrieved from cache.

We finally summarize in the following the three main conclusions we draw from the empirical results that are presented in the Appendix. A more analytical comment of the figures is postponed to the Appendix.

1. Our prefetching technique is highly beneficial to improve the performances of the system if integrated with a suitable caching algorithm. Greedy-Dual-Size with Prefetching (GDSP) outperforms on *hit rate* Least Recently Used with Prefetching (LRUP) and all the caching algorithms. The reason for this behaviour is that, differently from LRU, Greedy-Dual-Size tends to evict documents of small retrieval cost that can be easily prefetched later in the future. On the other end LRUP keeps in cache documents of large size at the expenses of the eviction of a large number of small documents. GDSP outperforms all other algorithms also in when compared on *reduced latency* and *byte-hit-rate*, even if this advantage is less remarkable than in terms of hit-rate. The reason for this is that popular documents have often smaller retrieval cost and size, while large documents, whose contribution to latency and network load is higher, are less frequently referenced in the sequence and therefore are weaker candidates for prefetching.

2. One-timers are never cached by our prefetching/caching algorithms. This results in a better use of the cache that allows GDSP, and partially also LRUP, to converge much faster than other algorithms. Due to the logarithmic shape of our metric of performances, this allows GDSP a dramatic saving of cache memory with respect to all other algorithms.

3. The advantage over pure Caching strategies increases on longer traces. This is very remarkable for instance in Figures 3, 4, 5, representing the hit rate on traces of $5,000-10,000$, $10,000-25,000$ and $25,000-50,000$ requests, where LRUP, the best of all other algorithms, achieves respectively about 80%, 50% and 25% of the hits of GDSP. This is not surprising since a more extensive information on the documents requested by the user later in the sequence allows to better exploit the potentiality of prefetching. This suggests that the performances of a similar system will improve together with its deployment on use.

References

1. Susanne Albers, Sanjeev Arora, and Sanjeev Khanna. Page replacement for general caching problems. In *Proceedings of the Tenth Annual ACM-SIAM Symposium on Discrete Algorithms*, pages 31–40, N.Y., January 17–19 1999. ACM-SIAM.

2. Susanne Albers, Naveen Garg, and Stefano Leonardi. Minimizing stall time in single and parallel disk systems. *JACM: Journal of the ACM*, 47, 2000.
3. Amotz Bar-Noy, Reuven Bar-Yehuda, Ari Freund, Joseph (Seffi) Naor, and Baruch Schieber. A unified approach to approximating resource allocation and scheduling. In *Proceedings of the Thirty-Second Annual ACM Symposium on the Theory of Computing*, Las Vegas, Nevada, 2000.
4. Azer Bestavros and Carlos Cunha. A prefetching protocol using client speculation for the world wide web. *Technical Report BUCS-TR-1995-011, Boston University, CS Dept, Boston, MA 02215*, April 1995.
5. P. Cao, E. Felten, A. Karlin, and L. Li. Implementation and performance of integrated application-controlled file caching, prefetching, and disk scheduling. *ACM Transactions on Computer Systems, 14(4)*, November 1996.
6. P. Cao and S. Irani. Cost-aware www proxy caching algorithms. *Proc. USENIX Symp. on Internet Technologies and Systems*, pages 193–206, 1997.
7. Dan Duchamp. Prefetching hyperlinks. *Proc. of the Second USENIX Symposium on Internet Technologies and Systems (USITS '99), Boulder, CO, USA*, 1999.
8. L. Fan, P. Cao, and Q. Jacobson. Web prefetching between low-bandwidth clients and proxies: Potential and performance. *Proc. of the Joint International Conference on Measurement and Modeling of Computer Systems (SIGMETRICS '99), Atlanta, GA*, May 1999.
9. J. Hine, C. Wills, A. Martel, and J. Sommers. Combining client knowledge and resource dependencies for improved world wide web performance. *Proc. of the INET '98 Conference, Geneva, Switzerland.Internet Society.*, July 1998. http://www.mcs.vuw.ac.nz/ hine/INET98paper/index.htm.
10. Ken ichi Chinen and Suguru Yamaguchi. An interactive prefetching proxy server for improvement of www latency. *Proc. of the Seventh Annual Conference of the Internet Society (INET'97), Kuala Lumpur*, July 1997.
11. T. Kimbrel and A. Karlin. Near-optimal parallel prefetching and caching. *Proc. 37th IEEE Annual Symposium on Foundations of Computer Science*, pages 540–549, 1996.
12. T. Kroeger, D. Long, and J. Mogul. Exploring the bounds of web latency reduction from caching and prefetching. *Proc. of the USENIX Symposium on Internet Technologies and Systems (USITS), Monterey, CA*, December 1997.
13. P. Lorenzetti, L. Rizzo, and L. Vicisano. Replacement policies for a proxy cache. *draft, http://www.iet.unipi.it/luigi/caching.ps.gz.*, 1996.
14. E. Markatos and C. Chronaki. A top-10 approach to prefetching on the web. *Proc. of the INET 98 Conference*, 1998.
15. V. Padmanabhan and J. Mogul. Using predictive prefetching to improve world wide web latency. *ACM SIGCOMM Computer Communication Review*, July 1996.
16. Z. Wang and J. Crowcroft. Prefetching in world wide web. *Proc. IEEE Global Internet 96*, November 1996.
17. N. Young. On-line file caching. *Proc. of the 9th Annual ACM-SIAM Symposium on Discrete Algorithms.*, pages 82–86, January 1998.

The Treewidth of Java Programs*

Jens Gustedt[1], Ole A. Mæhle[2], and Jan Arne Telle[2]

[1] LORIA & INRIA Lorraine, France
Jens.Gustedt@loria.fr
[2] University of Bergen, Norway
{olem,telle}@ii.uib.no

Abstract. Intuitively, the treewidth of a graph G measures how close G is to being a tree. The lower the treewidth, the faster we can solve various optimization problems on G, by dynamic programming along the tree structure. In the paper *M. Thorup, All Structured Programs have Small Tree-Width and Good Register Allocation* [8] it is shown that the control-flow graph of any `goto`-free C program is at most 6. This result opened for the possibility of applying the dynamic programming bounded treewidth algorithms to various compiler optimization tasks. In this paper we explore this possibility, in particular for Java programs.
We first show that even if Java does not have a `goto`, the labelled `break` and `continue` statements are in a sense equally bad, and can be used to construct Java programs that are arbitrarily hard to understand and optimize.
For Java programs lacking these labelled constructs Thorup's result for C still holds, and in the second part of the paper we analyze the treewidth of label-free Java programs empirically. We do this by means of a parser that computes a tree-decomposition of the control-flow graph of a given Java program. We report on experiments running the parser on several of the Java API packages, and the results tell us that on average the treewidth of the control-flow graph of these Java programs is no more than 2.7. This is the first empirical test of Thorup's result, and it confirms our suspicion that the upper bounds of treewidth 6, 5 and 4 are rarely met in practice, boding well for the application of treewidth to compiler optimization.

1 Background

Most structured language constructs such as `while`-loops, `for`-loops and `if-else` allow programs to be recursively decomposed into basic blocks with a single entry and `exit` point, see [1]. Such a decomposition corresponds to a series-parallel decomposition of the control-flow-graph of the program, see [6], and can ease static optimization tasks like register allocation, see [5]. On the other hand, with constructs such as the infamous `goto`, and also short-circuit evaluation of

---

* Part of this research was made possible by visiting funds from *The Norwegian Research Council* and the *Région Lorraine*

D. Mount and C. Stein (Eds.): ALENEX 2002, LNCS 2409, pp. 86–97, 2002.
© Springer-Verlag Berlin Heidelberg 2002

boolean expressions and multiple exit, break, continue, or return statements, this nice decomposition structure is ruined, see [5].

However, M. Thorup has shown in a recent article '*All Structured Programs have Small Tree-Width and Good Register Allocation*', see [8], that except for the goto, the other constructs listed above do allow for a related decomposition of the control-flow-graph of the program. For each of those language constructs, it was basically shown that regardless of how often they are used, they cannot increase the treewidth of the control-flow graph by more than one. Treewidth is a parameter that measures the 'treeness' of a graph, see [7]. Since a series-parallel graph has treewidth 2, this means that the control-flow-graphs of goto-free Algol and Pascal programs have treewidth ≤ 3 (add one for short-circuit evaluation), whereas goto-free C programs have treewidth ≤ 6 (add also for multiple exits and continues from loops and multiple returns from functions). Moreover, the related tree-decomposition is easily found while parsing the program, and this structural information can then, as with series-parallel graphs, be used to improve on the quality of the compiler optimization, see *e.g.* [2,3,8].

Informally, a graph of treewidth k can be decomposed by taking subsets of vertices of size ≤ $k + 1$, called bags, as nodes of a tree, in such a way that: 1) any edge of the graph has both endvertices in some bag, and 2) the bags containing any given vertex induce a connected subtree. Most NP-hard graph problems can be solved in linear time when restricted to a class of graphs for which there exists a constant k such that for each graph G in the class the treewidth of G is at most k. Such a linear-time algorithm on G will proceed by dynamic programming on its related width-k tree-decomposition, see [9]. The smaller the value of k, the faster the algorithm, so for the small bounds mentioned above, these algorithms may be quite useful. With unrestricted use of gotos one can for any value of k write a program whose control-flow graph has treewidth greater than k, so that the corresponding class of graphs does not have bounded treewidth. These results seem to imply that gotos are harmful for static analysis tasks. Gotos were originally considered harmful for readability and understanding of programs, see Dijkstra's famous article [4], and languages like Modula-2 and Java have indeed banned their use. Modula-2 instead provides the programmer with multiple exits from loops and multiple returns from functions with the pleasant consequence that all control-flow-graphs of Modula-2 programs have treewidth ≤ 5. In the paper by Thorup [8], the above-mentioned bounds on the treewidth of goto-free Algol, Pascal, C and Modula-2 are all given, but no mention is made of Java.

For a complete proof of the connection between treewidth and structured programs, we refer to [8]. However, by implementing Thorup's result in a Java parser we have gained a good intuition for this connection, which we summarize in the following paragraph.

The fact that control-flow graphs of programs that do not contain any Flow-Affecting Constructs (FACs) such as goto/break/continue/return and short-circuit evaluation of boolean expressions are series-parallel, and hence have treewidth 2, is based on the fact that all programming constructs/blocks of such programs have a single entry point and a single exit point, corresponding with the 2 terminals in the series and parallel operations. Also we know that allowing goto's all hope is lost, the control-flow graph can have arbitrarily high treewidth. However, for each of the 'label-free' FACs, continue/break/return and short-circuit evaluation, one can show that they respect the nesting structure of the program and also that there exists a tree-decomposition respecting this nesting structure. We take continue as an example. The explanation for the other label-free FACs will be very similar. A continue statement is related to a certain loop-block (while-loop or for-loop) of a program, and the target of this continue is the first statement of the loop-block. Any other continue related to this loop-block has the same target. All the statements of a loop-block B, including the target of a continue, can be found in the bags of a subtree T_B, and for a nested loop-block N inside of B again we find a similar subtree T_N of T_B, such that the bags of T_N contain no other statements from B except those of N and targets of any continues related to N. Thus, to show Thorup's result it suffices to take the width-2 tree-decomposition whose subtree structure reflects the nesting structure and for each label-free FAC added to the language, simply expand this tree-decomposition by, for each block B, taking the statement which is the target of the FAC at the outer level of block B, and adding it to every bag of T_B that does not belong to an inner nested block. See Figure 2 for an example, as produced by our parser implementing Thorup's result. We thus increase the treewidth by 1 for each label-free FAC added to the language, regardless of how many times each one is used in a program.

Obviously the introduction of labels make things more complicated in that regard, since each label introduces an extra jump target for both break and continue statements. In Section 2 we will prove technically that by introducing labels even in the restricted form as for Java with break and continue statements, the corresponding class of control-flow graphs does not have bounded treewidth, and things are in this sense as bad as if gotos were allowed. This is in contrast to the introduction of case-labels for which it is well known that they don't augment the treewidth by more than one, see [8]. It is also clear that the programming technique of the example we give can be misused to construct Java code that is arbitrarily challenging to understand.

In Section 3 we present our findings on the empirical study of the treewidth of label-free Java programs. In summary, the results are positive, showing an average treewidth of only 2.5. We also discuss the programming examples that have higher treewidth.

2   Not All Java Programs Are Structured

As compensation for the lack of a goto, the designers of Java decided to add the *labelled* break and continue statements. The latter two allow labelling of loops and subsequent jumping out to any prelabelled level of a nested loop. Java also contains exception handling, but we don't take this into account here. The main reason being that the interplay between optimization and exception handling (not only for Java) is quite unclear. On the compiler builder's side, the specification of Java doesn't tell too much about the actual implementation that is expected for exception handling, nor is there any emphasis on performance of the resulting code. Thus, a compiler optimization task like register allocation would apply only to exception-free execution of methods, executing exception-handling without any preallocation of registers.

In the original *'Go To Statement Considered Harmful'*-article, [4], what was in fact specifically objected to by Dijkstra was the proliferation of labels that indicate the target of `gotos`, rather than the `gotos` themselves. In fact, based on the results in this section, that article could aptly have been titled 'Labels Considered Harmful'. We show that, for any value of k, using only k labels, we can construct a Java program whose control-flow graph has treewidth $\geq 2k + 1$.

We will view the edges of the control-flow-graph as being undirected. Contracting an edge uv of a graph simply means deleting the endpoints u and v from the graph and introducing a new node whose neighbors are the union of the neighbors of u and v. A graph containing a subgraph that can be contracted to a complete graph on k nodes is said to have a clique minor of size k, and is well-known to have treewidth at least $k - 1$, see [7].

The labelled `break` and `continue` statements in Java allows the programmer to label a loop and then make a jump from a loop nested inside the labelled loop. In the case of a `continue` the jump is made to the beginning of the labelled loop, and in the case of a `break` the jump is made to the statement following the labelled loop. In the right-hand side of Figure 1 we show a listing of part of a Java program, with labels l1, l2 and l3, whose control-flow-graph can be contracted to a clique on 8 nodes.

```
continue1    l1:while (maybe) {
continue2    l2:  while (maybe) {break l1;
continue3    l3:    while (maybe) {break l1; break l2; continue l1;
innerloop            while (maybe) {break l1; break l2; break l3;
innerloop                            continue l1; continue l2; }
remainder3           break l1; break l2; break l3; continue l1; continue l2;}
break3             break l1; break l2; continue l1;}
break2           break l1;}
break1
```

Fig. 1. Skeleton of a Java program whose control-flow graph has treewidth ≥ 7. `Break` and `continue` statements should be conditional, but for the sake of simplicity this has been left out. The left column, in bold font, gives the names of contracted nodes of the control-flow-graph.

For simplicity we have chosen this code fragment that is obviously not real-life code, though it could easily be augmented to become more natural. For example, `breaks` and `continues` could be `case` statements of a `switch`.

Each of the 8 contracted nodes will naturally correspond to some lines of the corresponding Java program. Each of the 3 first lines of the listed code correspond to a node called, respectively, *continue1, continue2* and *continue3*, since they form the targets of the respective `continue` statements labelled l1, l2 and l3. The 4th and 5th lines of the code together form a node that we call *innerloop*, whereas the 6th line we call *remainder3* as it forms the remainder of the loop labelled l3. Lines 7 and 8 of the listing correspond to nodes that we call *break3* and *break2*, respectively, as they form the target of the `break` statements with labels l3 and l2. The target of the `break` labelled l1 is whatever statement that follows the listed code and it will be called *break1*, forming the eighth node.

It should be clear that each of these 8 nodes are obtained by contracting a connected subgraph of the control-flow-graph of the program. We now show that they form a clique after contraction, by looking at them in the order *innerloop, remainder3, continue3, break3, continue2, break2, continue1, break1* and arguing that each of them is connected to all the ones following it in the given order. Firstly, the node *innerloop* is connected to all the other nodes, as the control flows from it into *remainder3* when its loop entry condition evaluates to false, control flows naturally into *innerloop* from *continue3* and for each of the other 5 nodes *innerloop* contains the labelled `break` or continue statement targeting that node. Next, *remainder3* is connected to *continue3* as this is the natural flow of control, and *remainder3* contains the labelled `break` or `continue` statement targeting each of the other 5 nodes following it in the given order. The argument for the remaining nodes follows a similar line of reasoning. Morever, in the same style a larger code example can be made consisting of a method with k labels, a loop nesting depth of $k+1$ and a clique minor of size $2k+2$. The program lines following the line labelled l_k will for this larger example be:

```
lk: while (maybe) {break l1; ... break lk-1; continue l1; ... continue lk-2;
        while (maybe) {break l1; ... break lk; continue l1; ... continue lk-1; }
        break l1;...;break lk; continue l1;... continue lk-1;}
    break l1; ... break lk-1; continue l1; ... continue lk-2; }
```

Theorem 1 *For any value of $k \geq 0$ there exists a Java method with k labels and nesting depth $k-1$ whose control-flow-graph has treewidth $\geq 2k+1$.*

3   Treewidth of Actual Java Programs

If we restrict our focus to Java programs without labels, what can we say about the treewidth? The flow-affecting constructs available in Java are the same as those in C, with the exception that Java does not support the use of the `goto` statement. Thus from Thorup [8] we get the theoretical result that no control-flow graph of such programs have treewidth higher than 6. For a given Java method to achieve this high bound, it must contain, in addition to short-circuit evaluation, the flow-affecting constructs `break`, `continue` and `return`. However this is far from sufficient; for the width of the tree-decomposition to be raised by one for each of the constructs they need to "interfere" in the tree-decomposition, i.e. a bag in the decomposition must be affected by all the abovementioned constructs. This gives rise to the natural question of what we can *expect* the treewidth of a given Java method to be. To answer this question we implement a parser that takes as input programs written in Java, computes the corresponding tree-decomposition by Thorup's technique and thereby finds an upper bound on the treewidth of the control-flow graph.

3.1   Treewidth of Java API Packages

The classes of the API are organized in Java packages such as *java.io* and *java.util*. Thus the API is analyzed package-wise. Results from the tests are summarized in Table 1. The four rightmost coloumns shows the percentage-wise distribution of the methods with regard to treewidth, rounded to the nearest integer (except for values below 1, which is rounded to the nearest decimal). For example, package *java.lang* contains 604 methods. 27% of the methods have treewidth ≤ 2, 73% have treewidth 3, while only 1% may have treewidth as high as 4. The treewidth values computed by the parser is an *upper bound* on the treewidth of the control-flow graph of the methods, but we expect this bound to be tight in almost all cases.

Table 1. Treewidth of Java API packages

Package Name	# Methods	Avg. Treewidth	% tw 2	% tw 3	% tw 4	% tw 5
java.lang	604	2.73	27	73	1	0
java.lang.reflect	50	2.86	14	86	0	0
java.math	96	2.94	7	90	3	0
java.net	279	2.72	31	66	3	1
java.io	620	2.56	47	49	4	0
java.util	990	2.68	32	68	1	0
java.util.jar	93	2.73	28	71	1	0
java.util.zip	157	2.55	45	55	0	0
java.awt	1411	2.66	34	65	1	0
java.awt.event	71	2.74	25	75	0	0
java.awt.geom	527	2.71	30	69	1	0
java.awt.image	623	2.69	30	70	1	0
javax.swing	3400	2.62	39	60	1	0
javax.swing.event	87	2.63	37	63	0	0
javax.swing.tree	379	2.65	35	64	1	0
Total:	9387	Tot. Avg: 2.7				

Table 2. Treewidth of Java application programs

Application Name	# Methods	Avg. Treewidth	% tw 2	% tw 3	% tw 4	% tw 5
MAW D&A	391	2.51	49	51	0	0
MAW D&P	458	2.48	52	47	1	0
JAMPACK	260	2.6	41	58	1	0
Linpack	13	2.69	38	54	8	0
JIU	1001	2.53	48	51	1	0
Scimark2	57	2.59	40	60	0	0
JDSL	955	2.67	307	648	0	0
Total:	3135	Tot. Avg: 2.58				

Table 3. Brief description of tested Java applications

Application Name	Developer	Brief Description
MAW D&A	Mark Allen Weiss	Datastructures/Algorithms
MAW D&P	Mark Allen Weiss	Datastructures/Problem Solving
JAMPACK	G.W. Stewart	JAva Matrix PACKage
Linpack	Jack Dongarra, et.al.	Numerical Computation
JIU	Marco Schmidt	Image processing
Scimark2	Roldan Pozo et.al.	FFT ++
JDSL	Goodrich, Tamassia, et.al.	Data Structures Library

While the package test results varies some, the average Java API package typically has a distribution of the methods as follows. 20-40% of the methods have treewidth 2 and 60-80 % have treewidth 3. Only rarely are there more than 2% of the methods that cannot be guaranteed to have treewidth ≤ 3. Surprisingly, only one of the tested Java API packages have methods of treewidth 5, which is the *java.net* package. As it turns out this is only one method, namely *receive(DatagramPacket)* of class *java.net.DataSocket*. A closer look at this method is taken in Section 3.4.

3.2 Treewidth of Java Application Programs

Next we analyze the treewidth of ordinary Java applications. The programs were mostly found on the internet via search engines like Google (www.google.com). The bounds found are similar to those of the Java API classes. As Table 2 shows the results are similar to those of the Java API. Table 3 displays a short description of the applications. The choice of Java packages tested was based on availability of the source code.

3.3   Commonly Used Flow-Affecting Constructs

The 4 flow-affecting constructs (FACs) of Java are **break, continue, return** and short-circuit evaluation. We know that the treewidth may not necessarily increase by more than one even though several flow-affecting constructs are used within the same method. For instance a method containing **break, continue** and **return** may perfectly well be of treewidth 3. In fact, a program applying short-circuit evaluation may still have treewidth 2. This section examines what kind of flow-affecting structures are most widely used, and also to what extent flow-affecting constructs are used without increasing treewidth.

First we determine for the Java API packages of Section 3.1 how many of the methods use zero, one, two, three or four of the constructs in question, see Table 4. The first thing we observe is that the first columns of Tables 1 and 4 are almost identical. This is expected; the methods that don't utilize any of the flow-affecting constructs have treewidth 2. The differences between the two tables come from the cases where short-circuit evaluation is used without increasing treewidth.

Table 4. Flow-Affecting statements usage

Name	% using 0 FACs	% using 1 FACs	% using 2 FACs	% using 3 FACs	% using 4 FACs
java.lang	25	62	11	1	0.3
java.lang.reflect	14	80	6	0	0
java.math	7	76	17	0	0
java.net	31	57	11	2	0
java.io	46	41	10	3	0
java.util	31	58	10	1	0.1
java.util.jar	25	59	15	1	0
java.util.zip	45	51	4	0.6	0
java.awt	34	58	7	0.7	0
java.awt.event	25	55	18	1	0
java.awt.geom	27	60	13	0.6	0
java.awt.image	30	57	11	1	0
javax.swing	39	55	6	0.4	0
javax.swing.event	33	60	7	0	0
javax.swing.tree	35	54	11	0.3	0
Total Avg.	29.8	58.9	10.5	0.8	0.03

Looking at Table 1 almost all of the methods have treewidth 2 or 3. Comparing this to Table 4 we see that a number of the methods of treewidth 3 are split between having 1 or 2 flow-affecting constructs. In other words, methods commonly have 2 FACs, but treewidth ≤ 3. This is the case for about 10 percent of the methods in the Java API packages.

Next we analyze specifically what kind of flow-affecting constructs are most commonly used. We begin with the smallest of the applications, *Linpack*, for which we will give a somewhat more detailed description than the rest. Since the program doesn't have more than 13 methods we present all of them in Table 5. Throughout the program neither `break` nor `continue` are used at all, bounding the treewidth to 4. This corresponds nicely to our previous analysis of *Linpack*; one method having treewidth 4, the rest 2 or 3. Furthermore we observe, as expected, that the methods that have no flow-affecting contructs at all are exactly those of treewidth 2, while those that use one FAC are a subset of the methods of treewidth 3. Again we see that utilizing more than one flow-affecting construct doesn't necessarily increase treewidth by more than one, as is the case for method *ddot()*, which has 2 FACs, but treewidth 3.

Table 5. Flow-Affecting statements used in the *Linpack* program

Method Name	tw	FACs
main()	2	
abs()	3	return
second()	3	return
run_benchmark()	2	
matgen()	3	return
dgefa()	3	return
dgesl()	2	
daxpy()	4	short-sircuit, return
ddot()	3	short-circuit, return
dscal()	2	
idamax()	3	return
epslon()	3	return
dmxpy()	2	

Presenting data in the same manner as Table 5 from each of the methods of the Java API packages would hardly be suitable. Instead Table 6 shows how often the various flow-affecting constructs are used. We can see that by far the most widely used construct is the `return` statement, which is used by 65.5% of the methods in the Java API. Next, used by 13.2%, follows short-circuit evaluation, whereas `break` and `continue` is only found in 3.6 and 0.3% of the methods, respectively. The last coloumn shows how often the labelled `break`/`continue` statements are found. We see that 11 out of 15 packages do not use them at all, while *java.lang* contain either a labelled `break` or `continue` in 1% of the methods.

Table 6. Flow-affecting statements used in Java API packages

Name	% using return	% using break	% using continue	% using Short-Circuit	% using Labelled break/ continue
java.lang	72	2	1	14	1
java.lang.reflect	86	0	0	6	0
java.math	90	0	0	20	0
java.net	64	3	0.4	16	0
java.io	50	4	0.6	15	0.5
java.util	67	2	0.3	12	0.2
java.util.jar	69	3	1	19	0
java.util.zip	43	4	0	13	0
java.awt	60	3	0.2	12	0
java.awt.event	72	18	0	6	0
java.awt.geom	71	2	0.2	14	0.2
java.awt.image	65	7	0.2	11	0
javax.swing	55	2	0.2	11	0
javax.swing.event	64	3	0	7	0
javax.swing.tree	54	1	0	22	0
Total Avg.	65.5	3.6	0.3	13.2	0.13

3.4 A Java API Method of Treewidth 5

As previously mentioned, only one method was found to have treewidth 5. This was method *receive(DatagramPacket)* found in class *java.net.DataSocket*. It is therefore worth to take a closer look at this particular method, and see if we can decide why it achieves such a high bound. The relevant parts of the method are given in Figure 2, together with the generated tree-decomposition.

The excerpt consists of a `while` statement containing an if-else statement in which the expression utilizes short-circuiting. In addition to that, the `then` block of the expression has a `continue` and the `else` block has a `break` statement. As we know, each of these constructs can increase treewidth by one. Since they are all used within the same statement there exists a bag for which each of these constructs will increase the width, for a total width of 5, accounted for by the node with 3 children, which has a bag of size 6 in Figure 2.

4 Conclusion

Originally Java was designed to be precompiled to bytecode for the *Java Virtual Machine*, so compiler optimization tasks were then not a main issue. Nevertheless, since `gotos` were considered particularly harmful for the conceptual clarity of a program they were completely banned from the specification of Java, and a

Method Excerpt: **Tree–decomposition:**

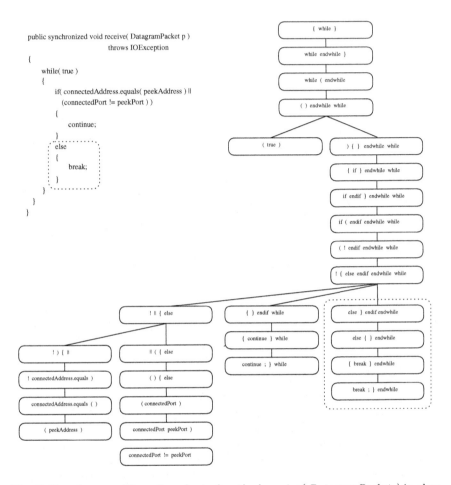

Fig. 2. Tree-decomposition of a subset of method *receive(DatagramPacket)* in class *DatagramSocket* of package *java.net*. Note how the indentation blocks (reflecting the scope depth) of the program appear as subtrees in the decomposition. For instance the *else{ break; }* block corresponds to the subtree pointed out by the dashed lines in the tree-decomposition.

labelled **break** and **continue** were added. Nowadays, to speed up applications written in Java, there is a strong demand for compiled *and* optimized Java, and so Java-to-native-machine-code compilers are emerging. In this paper we have shown that such compilers must have certain limits that are already inherent in the language itself. Nevertheless, programs that do not utilize labelled `break/continue` statements are of low treewidth on average. The experimental results of this paper justifies further research on how the tree-structure of these control-flow graphs can be utilized to improve various algorithms for compiler optimization tasks like register allocation.

References

[1] A. V. Aho, R. Sethi, and J. D. Ullman. *Compilers, Principles, Techniques and Tools.* Addison-Wesley, 1986.

[2] S. Alstrup, P. Lauridsen, and M. Thorup. Generalized dominators for structured programs. In *Proceedings of the 3rd Static Analysis Symposium*, volume 1145 of *LNCS*, pages 42–51, 1996.

[3] H. Bodlaender, J. Gustedt, and J. A. Telle. Linear-time register allocation for a fixed number of registers and no stack variables. In *Proceedings 9th ACM-SIAM Symposium on Discrete Algorithms (SODA'98)*, pages 574–583, 1998.

[4] E. W. Dijkstra. Go to statement considered harmful. *Comm. ACM*, 11(3):147–148, 1968.

[5] S. Kannan and T. Proebsting. Register allocation in structured programs. In *Proceedings of the Sixth Annual ACM-SIAM Symposium on Discrete Algorithms*, pages 360–368, San Francisco, California, 22–24 Jan. 1995.

[6] T. Nishizeki, K. Takamizawa, and N. Saito. Algorithms for detecting series-parallel graphs and *D*-charts. *Trans. Inst. Elect. Commun. Eng. Japan*, 59(3):259–260, 1976.

[7] N. Robertson and P. Seymour. Graph minors II. Algorithmic aspects of tree-width. *Journal of Algorithms*, 1986.

[8] M. Thorup. Structured programs have small tree-width and good register allocation. *Information and Computation*, 142(2), 1998.

[9] J. van Leeuwen. *Handbook of Theoretical Computer Science*, volume A, chapter Graph Algorithms - Classes of graphs, pages 545–551. Elsevier, Amsterdam, 1990.

Partitioning Planar Graphs with Costs and Weights[*]

Lyudmil Aleksandrov[1,3], Hristo Djidjev[2], Hua Guo[3], and Anil Maheshwari[3]

[1] Bulgarian Academy of Sciences, Sofia, Bulgaria
[2] Dept. of Computer Science, University of Warwick, Coventry, UK
hristo@dcs.warwick.ac.uk
[3] School of Computer Science, Carleton University, Ottawa, Canada
{hguo2,lyudmil,maheshwa}@scs.carleton.ca

Abstract. A graph separator is a set of vertices or edges whose removal divides an input graph into components of bounded size. This paper describes new algorithms for computing separators in planar graphs as well as techniques that can be used to speed up their implementation and improve the partition quality. In particular, we consider planar graphs with costs and weights on the vertices, where weights are used to estimate the sizes of the components and costs are used to estimate the size of the separator. We show that one can find a small separator that divides the graph into components of bounded size. We describe implementations of the partitioning algorithms and discuss results of our experiments.

1 Introduction

Motivation: Graph partitioning is a basic tool in the design of efficient algorithms for solving algorithmic as well as application problems. Informally, the problem consists of finding, in an input graph G, a small set of vertices or edges, called a *separator*, whose removal divides G into two or more components of roughly equal sizes. A variety of algorithms for solving combinatorial problems are based on the divide-and-conquer strategy combined with an efficient use of graph partitioning on planar graphs. Such algorithms include the best known solutions to the shortest path problem for planar or near-planar graphs, divide-and-conquer algorithms in combinatorial optimization and computational geometry, efficient parallel factorization of sparse matrices, and algorithms for finding a VLSI layout design. In scientific computing, graph partitioning is used to find a distribution of the data among the processors in a high performance computing system that balances the load and minimizes the interprocessor communication.

The goal of this work is to develop and do experiments with techniques that increase the speed of graph partitioning algorithms and/or improve the quality of the partitioning. All of our algorithms have theoretically guaranteed upper

[*] Research supported in part by NSERC and NCE GEOIDE. Second author was partially supported by the EPSRC grant GR/M60750, RTDF grant 98/99-0140, and IST Programme of the EU, Contract IST-1999-14186, ALCOM-FT.

D. Mount and C. Stein (Eds.): ALENEX 2002, LNCS 2409, pp. 98–110, 2002.
© Springer-Verlag Berlin Heidelberg 2002

bounds both on the quality of the partitions and on the time needed to find them that are optimal for the class of planar graphs within a constant factor.

Related Work: Lipton and Tarjan were first to formally study the graph partitioning problem in [10], where they introduced the notion of a *graph separator*. They showed that in any n-vertex planar graph with nonnegative vertex weights summing to $w(G)$ there exists a separator of no more than $\sqrt{8n}$ vertices whose removal divides the graph into components none of whose weights is more than $(2/3)w(G)$. Generalizing the notion of the size of a graph separator, Djidjev [4] considers planar graphs with real-valued positive costs and weights on the vertices. The main theorem in [4] states that in a planar graph $G = (V, E)$ there exists a vertex set of total cost $2\sqrt{2\sigma(G)}$, where $\sigma(G) = \sum_{v \in V}(c(v))^2$, whose removal leaves no component of weight exceeding $(2/3)w(G)$, where $c(v)$ denotes the cost of a vertex v. Moreover, it is shown that this bound is tight up to a constant factor and that the partition can be computed in linear time.

In spite of the extensive algorithmic work on constructing separators or using them in the design of other combinatorial algorithms, there is very little work on the efficient implementation of the new partitioning algorithms. Gilbert et al. describe in [6] a Matlab implementation of a partitioning algorithm for geometric graphs (graphs whose embedding in two or three dimensions satisfies certain requirements) based on a theoretical work of Miller, Teng, Thurston, and Vavasis. A number of researchers, mainly working in the scientific computing area, have designed and implemented graph partitioning algorithms which are based on heuristics like, e.g., the Kernighan-Lin method [9], eigenvectors of graphs [13], genetic algorithms [12], or combinations of different methods (see [2] for a survey of results). Although these algorithms usually work well in practice for the chosen applications, they can not guarantee a balanced partition in the worst case. Software packages containing implementations of such type of algorithms include Chaco [7] and METIS [14].

New results: Let $G = (V, E)$ be an n-vertex planar graph with real-valued positive weights $w(v)$ and costs $c(v)$ associated with its vertices. Let t be a real in $(0, 1)$. A set of vertices (edges) S is called a vertex (edge) t-*separator* if its removal leaves no component of weight exceeding $tw(G)$. Main results of this paper are

1. We show that there exists a vertex t-separator of G whose cost is at most $4\sqrt{2\sigma(G)/t}$, where $\sigma(G) = \sum_{v \in V}(c(v))^2$, and provide an algorithm for constructing such a separator. The cost of the separator is asymptotically optimal for the class of planar graphs. The running time of the algorithm is $O(n + T_{\mathrm{SSSP}}(G))$, where $T_{\mathrm{SSSP}}(G)$ is the time for computing single source shortest path tree from a fixed vertex in G[1].

2. We present a technique for tuning above algorithm that results in considerable reduction of the cost of the produced separators. The computational

---

[1] In Klein et al. [8] it is shown that SSSP tree in planar graphs can be computed in linear time, but the algorithm is complicated and the hidden constant seems to be large. Alternatively, we use Dijkstra's algorithm and obtain simple and practical algorithm.

cost of this technique is analyzed theoretically in the case of integer costs
and experimentally tested in the general case.

3. We show that there exists an edge t-separator of at most $4\sqrt{2\Delta(G)/t}$ edges,
   where $\Delta(G) = \sum_{v \in V}(\deg(v))^2$ and $\deg(v)$ is the degree of a vertex v. The
   size of the edge separator is asymptotically optimal for the class of planar
   graphs. Our algorithm constructs such a separator in $O(n + T_{\mathrm{SSSP}}(G))$ time.
4. We implement our algorithms and support the theoretical results by extensive experiments. Unlike other implementations of graph partitioning algorithms based on heuristics, ours always produce balanced partitions and
   separators whose worst-case cost (size) is guaranteed. Moreover, the experiments suggest that the constants in the estimates on the cost of separators
   can be improved. In addition, experiments indicate that our tuning technique
   is effective and computationally efficient.

Result 1 is a generalization of many existing results on separators, which follow
by appropriately choosing weights, costs, and a parameter t. In most of the previous work on separators, the parameter t is chosen to be $2/3$ and a fast (usually
linear time) algorithm is presented to obtain such a partition. Partitions with
smaller values of t have been constructed, using $2/3$-separators, by recursively
applying the corresponding algorithm, which adds an extra $O(\log(1/t))$ factor
to time complexity. In comparison, our Result 1 works for any value of t and
hence t-separators are constructed in linear time for any t. This is supported
by our experiments Moreover, our experiments show that the execution time is
consistently small for different values of t and hence our results are not only
theoretically, but also practically attractive.

Separators for graphs with vertex weights and costs have applications in
efficiently solving combinatorial problems such as embedding of planar graphs
into communication structures, graph pebbling , and construction of a minimum
routing tree. See [4] for further discussion of these and other problems.

2   Partitioning of Planar Graphs with Weights and Costs

Outline of our approach: Our algorithm is based on the approach introduced
in [10] and extended in [1,4]. In this approach, construction of the separator relies
on a SSSP tree T rooted at a dummy vertex ρ of zero weight and cost that is
added to G and connected to at least one of its vertices (so that planarity is
preserved). The distance in G is related to the cost of vertices. That is, the cost
of traveling along a single edge equals the cost of its target vertex and the cost
of a path is the sum of the costs of the edges in that path. In the uniform case,
where all vertices have equal costs, T is simply the breadth first search tree
rooted at ρ.

The algorithm constructs the separator in three phases. In the first phase,
a set of levels in T are selected and vertices in these levels are added to the
separator. Intuitively, a level in T consists of the vertices that are at the same
distance from ρ and its removal partitions the graph into two disjoint subgraphs
induced by the vertices "below" and "above" that level. After the completion of

the first phase, the removal of the levels in the current separator partitions G into components $G_0, \ldots, G_p$ with relatively "short" SSSP trees.

In the second phase, each connected component G_j whose weight exceeds $tw(G)$ is further partitioned as follows. First, a dummy vertex ρ_j of zero weight and zero cost is added, then the component is triangulated and an SSSP tree T_j *consistent* with T is constructed. By consistent we mean that the distance between any two vertices in T_j is at most their distance in T. Then G_j is partitioned by means of fundamental cycles with respect to T_j. Recall that a fundamental cycle is a cycle consisting of a single non-tree edge plus the tree path between its endpoints. We find a low cost set of fundamental cycles whose removal partitions G_j into components of weights at most $tw(G)$ by applying the so called *separation tree* technique, which is described in details in [1]. Vertices in this set of fundamental cycles are added to the separator. After the second phase we obtain a set of vertices S_t in the separator whose removal leaves no component of weight bigger than $tw(G)$ and whose cost satisfies the inequality

$$c(S_t) = \sum_{v \in S_t} c(v) \leq 4\sqrt{2\sigma(G)/t}. \tag{1}$$

In the third phase, called *packing*, we "clean up" the separator by removing vertices that are not adjacent to at least two different components as well as by merging neighboring components of small weight into a new component provided that the weight of the new one is still smaller than $tw(G)$.

Construction of a t-separator: We first discuss the second phase of the algorithm. Let $G = (V, E)$ be a triangulated planar graph, where each vertex $v \in V$ has weight $w(v)$ and cost $c(v)$. Let T be a spanning tree of G rooted at a vertex ρ. Define the radius of T to be the cost of the longest path in T originating at ρ and denote it by $r(T)$. In [1] it is shown that there exists a set of fundamental cycles $C_1, C_2, \ldots, C_q$ with $q \leq \lfloor 2/t \rfloor$ whose removal leaves no component with weight exceeding $tw(G)$. Denote the set of vertices in these cycles by C. By the definition C is a t-separator and its cost can be estimated by $c(C) = \sum_{v \in C} c(v) \leq \sum_{i=1}^{q} c(C_i) \leq 2qr(T) \leq 2\lfloor 2/t \rfloor r(T)$, since any fundamental cycle in T has cost at most $2r(T)$.

Theorem 1. *Let G be a triangulated planar graph with positive weights and costs on vertices. Let T be a spanning tree of G with radius $r(T)$. There exists a set C of fundamental cycles forming a t-separator such that its cost does not exceed $2\lfloor 2/t \rfloor r(T)$.*

Next, we describe the first phase of our algorithm, that partitions G into components with "short" SSSP trees by using levels. Assume for simplicity, that our graph G is connected and let T be a SSSP tree rooted at ρ. Let vertices of G be enumerated by their distance to ρ, i.e., $0 = d(\rho) < d(v_1) \leq \ldots \leq d(v_n) = r(T)$, where n is the number of vertices in G and $d(v_i)$ denotes the distance from ρ to v_i.

Definition 1. *For any real $x \in [0, r(T)]$ we define a set of edges $E(x)$ by $E(x) = \{e = (u, v) : e \in E(T), \ d(u) < x \leq d(v)\}$ and a set of vertices $L(x)$ called* level *by $L(x) = \{v : e = (u, v) \in E(x), \ d(u) < d(v)\}$.*

The following three lemmas can be easily established from the definition.

Lemma 1. *Let x, y be reals in $[0, r(T)]$. If the interval (x, y) does not contain any of the distances $d(v_i)$, then $L(x) = L(y)$.*

Hence, there are at most n different levels in G, i.e., for any $x \in [0, r(T)]$ the level $L(x)$ coincides with one of the levels $L(d(v_1)), \ldots, L(d(v_n))$. Any of the levels $L(x)$, $x \in (0, r(T))$, cuts G into two subgraphs as stated in the following lemma.

Lemma 2. *For any real $x \in [0, r(T)]$ the level $L(x)$ partitions vertices of G into two subsets $V_- = \{v : d(v) < x\}$, and $V_+ = \{v : d(v) > x, v \notin L(x)\}$, such that no edge in G joins V_- and V_+.*

Therefore, the removal of a level $L(x)$ with $x \in (0, r(T))$ partitions G into two subgraphs G_- and G_+ induced by the sets V_- and V_+, respectively. Next lemma computes the integral on $c(L(x))$, which appears in the estimate on the cost of our separators.

Lemma 3. $\int_0^{r(T)} c(L(x)) d(x) = \sigma(G) = \sum_{v \in V} (c(v))^2$.

Lemma 2 and Theorem 1 can be used together for obtaining a variety of t-separators in G. Namely, let $i_1, \ldots, i_p$ be a collection of increasing indices, i.e. $0 = i_0 < i_1 < \ldots < i_p < i_{p+1} = n + 1$. Removal of the levels $L(d(v_{i_1})), \ldots, L(d(v_{i_p}))$ partitions G into subgraphs $G_0, G_1, \ldots, G_p$. For each of the graphs G_j whose weight exceeds $tw(G)$ we construct a $tw(G)/w(G_j)$-separator S_j of fundamental cycles with respect to a spanning tree T_j consistent with T. In case that $w(G_j) \leq tw(G)$ we assume that $S_j = \emptyset$. Separators S_j are constructed as follows. Consider the graph G_0. We triangulate it and then find an SSSP tree T_0 rooted at ρ. Obviously, T_0 is consistent with T. Applying Theorem 1 we obtain a $tw(G)/w(G_0)$-separator S_0 such that $c(S_0) \leq 2\lfloor 2w(G_0)/tw(G)\rfloor r(T_0) \leq 2\lfloor 2w(G_0)/tw(G)\rfloor(d(v_{i_1-1}) - d(\rho))$. Next, consider the graph G_1. We connect ρ to all the vertices in G_1 adjacent to $L(d(v_{i_1}))$ in G. The resulting graph is planar. We triangulate that graph and find an SSSP tree T_1 rooted at ρ. The tree T_1 is consistent with T. Using Theorem 1 we construct a $tw(G)/w(G_1)$-separator S_1 with cost $c(S_1) \leq 2\lfloor 2w(G_1)/tw(G)\rfloor r(T_1) \leq 2\lfloor 2w(G_1)/tw(G)\rfloor(d(v_{i_2-1}) - d(v_{i_1}))$. In this way, for each $j = 0, 1, \ldots, p$, we obtain a set of vertices S_j that is $tw(G)/w(G_j)$-separator for G_j and such that

$$c(S_j) \leq 2\lfloor 2w(G_j)/tw(G)\rfloor(d(v_{i_{j+1}-1}) - d(v_{i_j})). \tag{2}$$

Clearly, the union of the levels $L(d(v_{i_1})), \ldots L(d(v_{i_p}))$ and separators $S_0, \ldots, S_p$ forms a t-separator for G. We denote that separator by $S(i_1, \ldots, i_p)$. Its cost is estimated by

$$c(S(i_1, \ldots, i_p)) = c\left(\bigcup_{j=1}^{p} L(d(v_{i_j}))\right) + c\left(\bigcup_{j=0}^{p} S_j\right) \leq \tag{3}$$

$$c\left(\bigcup_{j=1}^{p} L(d(v_{i_j}))\right) + 2\sum_{j=0}^{p} \left\lfloor \frac{2w(G_j)}{tw(G)} \right\rfloor (d(v_{i_{j+1}-1}) - d(v_{i_j})) = \Gamma(i_1, \ldots, i_p),$$

where i_{p+1} is assumed $n + 1$. In the next theorem we prove that there exists t-separator $S(i_1, \ldots, i_p)$ of small cost by an appropriate choice of the indices $i_1, \ldots i_p$.

Theorem 2. *Let G be an n-vertex planar graph with positive weights and costs associated to its vertices. For any $t \in (0,1)$ there exists a t-separator S_t such that $c(S_t) \leq 4\sqrt{2\sigma(G)/t}$. Separator S_t can be found in $O(n)$ time in addition to the time for computing an SSSP tree in G.*

Proof: Let $h = \sqrt{t\sigma(G)/8}$. For $j = 0, \ldots, p$ define equidistant points $y_j = jh$ in $[0, d(v_n)]$, where $p = \lfloor d(v_n)/h \rfloor$. Points $y_1, \ldots, y_p$ divide the interval $[0, d(v_n)]$ into $p + 1$ subintervals such that first p have length h and the last one has length less than h. For $j = 1, \ldots, p$, let x_j be a point in $[y_{j-1}, y_j]$ where $c(L(x_j))$ is minimum, i.e., we have $c(L(x_j)) \leq c(L(x))$ for $x \in [y_{j-1}, y_j]$. Further, let i_j be the smallest index such that $x_j \leq d(v_{i_j})$. By Lemma 1, we know that $L(x_j) = L(d(v_{i_j}))$. Consider now the t-separator $S(i_1, \ldots, i_p)$ constructed as described above. By (3), for the cost of this separator we have

$$c(S(i_1, \ldots, i_p)) \leq c\left(\bigcup_{j=1}^{p} L(d(v_{i_j}))\right) + 2\sum_{j=0}^{p} \left\lfloor \frac{2w(G_j)}{tw(G)} \right\rfloor (d(v_{i_{j+1}-1}) - d(v_{i_j})).$$

We estimate the two terms on the right side separately. For the first term we use the definition of the points x_j and Lemma 3

$$c\left(\bigcup_{j=1}^{p} L(d(v_{i_j}))\right) \leq \sum_{j=1}^{p} c(L(x_j)) \leq \sum_{j=1}^{p} \frac{1}{h} \int_{y_{j-1}}^{y_j} c(L(x))dx \leq \qquad (4)$$

$$\frac{1}{h} \int_{0}^{d(v_n)} c(L(x))dx = \sigma(G)/h.$$

To estimate the second term we observe that $d(v_{i_{j+1}-1}) - d(v_{i_j}) \leq 2h$ for each $j = 0, \ldots, p$ and thus

$$2\sum_{j=0}^{p} \left\lfloor \frac{2w(G_j)}{tw(G)} \right\rfloor (d(v_{i_{j+1}-1}) - d(v_{i_j})) \leq 4h \sum_{j=0}^{p} \left\lfloor \frac{2w(G_j)}{tw(G)} \right\rfloor \leq 8h/t \qquad (5)$$

To obtain the theorem we sum (4) and (5) and substitute h with its value $\sqrt{t\sigma(G)/8}$. The time bounds can be established easily. $\square$

Applications of Theorem 2: Theorem 2 in essence captures many of the existing results on separators in planar graphs. By appropriately setting the cost and weights of vertices in the given planar graph, we can obtain variety of separator results. This includes the vertex separator of Lipton and Tarjan [10], edge separators [3], bisectors [4], to name a few. Below we show how Theorem 2 can be applied to construct an edge t-separator in a weighted planar graph. Recall that a set of edges is called a *edge t-separator* of G, if their removal leaves no component of weight greater than $tw(G)$.

Theorem 3. *Let G be an n-vertex planar graph with positive weights on vertices. Let t be a real in $(0, 1)$ and let no vertex of G have weight exceeding $tw(G)$. There exists an edge t-separator of G consisting of at most $4\sqrt{2\Delta(G)/t}$ edges, where $\Delta(G) = \sum_{v \in V}(deg(v))^2$ and $deg(v)$ is the degree of v. Such a separator can be found in $O(n)$ time in addition to computing SSSP tree in G.*

Proof: First we set the cost of each vertex as its degree. Then we apply Theorem 2 and obtain a vertex t-separator VS_t whose cost is at most $4\sqrt{2\Delta(G)/t}$. Since the cost of a vertex is its degree, there are at most $4\sqrt{2\Delta(G)/t}$ edges incident to vertices in VS_t. Next, for each vertex $v \in VS_t$ we consider components of $G \setminus VS_t$ adjacent to v and if there is a component whose weight plus the weight of v is less than $tw(G)$ we absorb v in that component. If there is no such component we make v to constitute its own component. Finally, we form an edge t-separator ES_t by inserting in it all edges incident to vertices in VS_t that join different components. Clearly, ES_t satisfies the requirements of the theorem. $\square$

3   Reducing the Cost of the Separators

In this section we describe a technique that optimizes the choice of levels to be inserted in the separator during the first phase of the algorithm. Recall that in the first phase we select a collection of indices $i_1, \ldots, i_p$ and then cut the graph by removing levels $L(d(v_{i_1})), \ldots, L(d(v_{i_p}))$. We have shown that the cost of the separator $S(i_1, \ldots, i_p)$ resulting after the choice of levels corresponding to these indices is bounded by $\Gamma(i_1, \ldots, i_p)$ from (3). To prove our main result in the previous section (Theorem 2) we have shown how to choose a collection of indices such that the value of Γ does not exceed $4\sqrt{2\sigma(G)/t}$. As we have seen, these indices can be found efficiently in linear time and the cost of the resulting separators $\Gamma(i_1, \ldots, i_p)$ is asymptotically optimal; hence such a choice is satisfactory from asymptotic complexity point of view.

In practice, however, we may apply a simple and natural heuristic. That is, we may find a collection of indices for which the value of Γ is minimum and then use the corresponding levels to partition the graph. It is not possible to argue that such a choice is superior in all cases since we are minimizing a function that is an upper bound on the cost of the separator but not the cost itself. Nevertheless, our experiments (see Table 3) show that this heuristic significantly improves the cost of the separators and is computationally efficient. In the remainder of this section we describe an algorithm for finding a collection of indices $i_1, \ldots, i_p$ such that the value $\Gamma(i_1, \ldots, i_p)$ is minimum.

Reformulating the problem as a shortest path problem: Notice that the number of different levels is not always n. If two vertices have the same distance to the root ρ then their levels coincide. Assume that there are $n' \leq n$ vertices with distinct distances from ρ. Assume that vertices with the same distance are grouped and let for simplicity denote these groups with $v_1, \ldots, v_{n'}$ so that $0 = d(\rho) < d(v_1) < \ldots < d(v_{n'}) = r(T)$. Thus we have n' distinct levels $L(d(v_i))$, $i = 1, \ldots, n'$ to chose from. Consider now a graph $H = (V(H), E(H))$ defined by $V(H) = \{0, 1, \ldots, n' + 1\}$ and $E(H) = \{(i, j) : 0 \le i < j \le n' + 1\}$. We call H *levels graph* and assign costs to its edges so that the cost of a path from 0 to $n' + 1$ and passing through $i_1, \ldots i_p$ is exactly $\Gamma(i_1, \ldots, i_p)$. This obviously is achieved by defining the cost of an edge (i, j) as

$$cost(i, j) = c(L(v_j) \setminus L(v_i)) + 2\lfloor 2w(G_{i,j})/tw(G)\rfloor (d(v_{j-1}) - d(v_i)),$$

where $L(v_0) = L(v_{n'+1}) = \emptyset$ and $w(G_{i,j})$ is the total weight of the vertices with distances between $d(v_i)$ and $d(v_j)$ and not in $L(d(v_i))$. Now the problem of finding a collection of indices minimizing Γ can be solved by finding the shortest path between vertices 0 and $n' + 1$ in the level graph H.

Efficient computation of shortest paths in H: Dijkstra's shortest path algorithm can be used to find the shortest path between 0 and $n' + 1$ and hence a collection of indices for which Γ is minimized. The problem with direct application of Dijkstra's algorithm applied on H is that it takes $\theta(n'^2)$ time, which can be $\Omega(n^2)$. Below we discuss how the special structure of graph H can be exploited so that Dijkstra's algorithm will not need to explore all the edges of H. The classical Dijkstra's algorithm maintains a priority queue Q storing the vertices of the graph, where the key of a vertex i in Q is the shortest distance from 0 to i found so far. In each iteration the algorithm extracts the vertex with a minimum key from Q, which we refer to as a *current vertex*, and then the edges incident to the current vertex are *relaxed*. Next we describe briefly the main points in our efficient implementation of Dijkstra's algorithm on H.

Notice that we can not precompute the cost of the edges since this will take $\Omega(n'^2)$ time. But we can precompute the costs and weights of the sets of vertices $L(d(v_{i+1})) \setminus L(d(v_i))$ for $i = 1, \ldots, n'$ in $O(n)$ time. Using these precomputed values, we can compute the cost of an edge (i, j) in $O(1)$ time provided that we know the cost of $(i, j - 1)$ or $(i, j + 1)$. This leads to the following set of rules for our implementation.

Rule 1 *Edges incident to the current vertex are relaxed following their consecutive order (increasing or decreasing).*

For a vertex i of H denote by $dist(i)$ the cost of the shortest path from 0 to i.

Proposition 1. *If $i < j$ and $dist(i) \ge dist(j)$ then there is a shortest path from 0 to $n' + 1$ that does not include i.*

Rule 2 *The indices of the consecutive current vertices must increase.*

Furthermore, from Theorem 2 it follows that $dist(n' + 1) \le B = 4\sqrt{2\sigma(G)/t}$. Therefore we do not need to consider paths longer than B during the implementation of Dijkstra's algorithm. To employ this observation we notice that the cost of an edge (i, j) in H is a sum of two terms. We call them *cost part* and *weight part* and denote them by $cost_c(i, j)$ and $cost_w(i, j)$. We observe that the weight part of an edge is an increasing function with respect to the target vertex.

Rule 3 *If i is the current vertex and $dist(i) + cost_w(i, j) > B$, then edges (i, j') with $j' > j$ are not relaxed.*

Applying this rule during the relaxation we can discard all edges "after" some edge. Next lemma shows that we can discard all edges "before" some edge under certain conditions.

Proposition 2. *If $i' < i < j$ and $dist(i') + cost(i', j) \leq dist(i) + cost(i, j)$, then $dist(i') + cost(i', j') \leq dist(i) + cost(i, j')$ for any $i < j' < j$.*

Rule 4 *If i is the current vertex and the relaxation of an edge (i, j) does not result in smaller distance to j then edges (i, j') with $j' < j$ are not relaxed.*

The above four rules suggest the following implementation of Dijkstra's algorithm on H. For each vertex i we compute a vertex $jump(i)$ that is the smallest one such that $w(G_{i,jump(i)}) > tw(G)$. Assume that i is a current vertex whose edges are to be relaxed. We relax the edge $(i, jump(j))$ first. Then edges before $jump(i)$ are relaxed in decreasing order until the first unsuccessful relaxation is encountered (Rule 4). Next edges after $(i, jump(i))$ are relaxed in their increasing order until Rule 3 applies. Our experiments show that this results in a very efficient procedure and timings suggest that the running time is at most $O(n')$. So far, we do not have a rigorous proof of this bound, whereas some of the special cases are characterized in the following lemma

Lemma 4. *If the cost of vertices are positive integers a collection of indices minimizing Γ can be found in $O(\min(n^2, \sigma(G)/t))$ time. In the uniform case, when all vertices have the same cost, the time bound is $O(\min(n^2, n/t))$.*

4   Implementation and Experimental Results

In this section we provide relevant details on our implementation of vertex and edge separators and present supporting experimental results. The algorithms are implemented in C++ using LEDA [11] on Sun Ultra 10 Microsystems, 440MHz clock, 512 Mb memory running Solaris 7 operating system. For experiments we use planar graphs produced by LEDA graph generators as well as planar graphs available on the web[2], including finite element meshes, irregular planar graphs and triangular irregular networks. We have experimented on a large number of graphs, a sample of which is shown in the tables below. We have implemented the algorithm for finding vertex t-separators in planar graphs with weights and costs as described in Section 2 and using the cost reducing technique of Section 3. That is, in phase I a set of levels is determined using cost reducing technique of Section 3. Then in phase II each "heavy" component resulting after the removal of levels is further partitioned by using fundamental cycles. After phase I and II a t-separator is obtained whose cost is less than $4\sqrt{2\sigma(G)/t}$. Ideally, the number

---

[2] e.g., see
   http://www.uni-paderborn.de/fachbereich/AG/monien/RESEARCH/graphs.html

of components resulting after the removal of a t-separator should be about $1/t$. However partitions resulting after phase I and II usually consist of more than $1/t$ components due to a number of small weight components generated by the algorithm. Thereby in phase III, called packing, partition is further improved by merging small weight components and cleaning the separator from vertices that are not adjacent to two or more different components. Packing is more successful if the bound on the weight of the components is relaxed to $\alpha tw(G)$ with some $\alpha \geq 1$. Indeed there is a tradeoff between maximum allowed weight of a component and the cost (size) of the resulting separators. Results presented below were obtained with $\alpha = 1.5$. In Table 1 we present the results when the algorithm is run on graphs with unit weights and costs. Then in Table 2 we present the results when the algorithms is run on the same graphs, but with weights and costs assigned to the vertices. Weights and costs were randomly generated in the interval $[1, 100]$.

Table 1. Experimental results on partitioning graphs with n vertices and m edges and with unit weights and costs. Cost reducing technique is used. Time is in seconds and time for phase I + II and phase III is shown separately. During phase III upper bound on the size of a component was relaxed to $1.5tn$. Number of components as well as minimum and maximum size of a components are shown.

Graph	$\frac{1}{t}$	$\frac{|S|}{\sqrt{n/t}}$	$|S|$	# of comp.	min size	max size	time for Ph. I+II	time for Phase III
tapir	2	0.376	17	2	367	640	0.12	0.00
$n = 1024$	4	0.313	20	4	162	301	0.10	0.00
$m = 2846$	32	1.143	207	28	14	40	0.10	0.01
	128	1.057	383	87	3	11	0.08	0.01
airfoil1	2	0.434	40	2	1791	2422	0.55	0.03
$n = 4253$	4	0.889	116	4	897	1314	0.61	0.04
$m = 12289$	32	1.431	528	30	107	163	0.46	0.03
	128	1.511	1115	108	18	42	0.39	0.05
triangle	2	0.706	71	2	2485	2494	0.02	0.04
$n = 5050$	4	1.048	149	4	1176	1261	0.35	0.04
$m = 14850$	32	1.567	630	30	135	188	0.52	0.02
	128	1.620	1303	115	24	49	0.53	0.04
tin100000	2	0.710	226	2	24867	25531	9.08	0.48
$n = 50624$	4	0.998	449	4	12432	12656	6.81	0.47
$m = 150973$	32	1.637	2083	32	1025	1580	8.68	0.48
	128	1.766	4496	124	256	437	8.58	0.52

There are several observations that we can make from Tables 1 and 2. Costs of the obtained separators are significantly smaller than the upper bound derived in Theorem 2. This is illustrated by the ratio between the cost of the separator $c(S_t)$ and $\sqrt{\sigma(G)/t}$. By Theorem 2 this ratio is at most $4\sqrt{2} \approx 5.657$, whereas we never get this ratio bigger than 2 in our experiments. Overall running times are small and do not change significantly by altering parameter t. Times in the nonuniform case are slightly bigger due to the construction of SSSP tree and increased time for the cos reducing technique. This suggest that these algorithms are practical. Also, Phase III, which combines the small components into larger ones respecting the relaxed weight criteria, takes only a small fraction of time compared to the time spent in Phase I plus Phase II. These observations suggest that our algorithms are practical.

Table 2. Experimental results on partitioning graphs whit n vertices and m edges and with random weights and costs from the interval $[1, 100]$ assigned to vertices. Cost reducing technique is used. Time is in seconds and time for Phase I + II and Phase III is shown separately. During phase III upper bound on the size of a component was relaxed to $1.5tw(G)$. Number of components as well as minimum and maximum weight of a component are shown.

Graph	$\frac{1}{t}$	$\frac{c(S)}{\sqrt{\sigma(G)/t}}$	$c(S)$	$|S|$	# of comp.	min weight	max weight	time of Ph. I+II	time of Phase III
tapir	2	0.116	304	10	2	23837	28236	0.13	0.01
$n = 1024$	4	0.290	1079	33	4	11112	15108	0.15	0.00
$m = 2846$	32	0.898	8666	185	27	944	2440	0.15	0.01
	128	0.918	17307	367	103	81	576	0.12	0.02
airfoil1	2	0.447	2423	85	2	86158	125516	0.79	0.04
$n = 4253$	4	0.425	3263	114	4	39842	65029	0.76	0.03
$m = 12289$	32	1.062	23031	602	32	3961	8270	0.70	0.04
	128	1.448	57189	1234	106	898	2094	0.74	0.04
triangle	2	0.324	1901	76	2	92979	159126	0.82	0.03
$n = 5050$	4	0.720	5982	224	4	56783	70892	0.95	0.04
$m = 14850$	32	1.264	28741	771	30	5593	10336	0.86	0.04
	128	1.526	63380	1489	100	1104	2754	0.88	0.05
tin100000	2	0.345	6397	254	2	1195571	1335566	13.73	0.49
$n = 50624$	4	0.617	16166	664	4	572835	753388	10.04	0.48
$m = 150973$	32	1.288	99829	3082	35	54301	92659	14.32	0.48
	128	1.510	229003	6476	134	11130	25668	14.57	0.52

Next, we test the efficiency of cost reducing technique from Section 3. We compare two different implementations of Phase I of the t-separator algorithm. In the first version we implement the t-separator by choosing the levels as described in the proof of Theorem 2. We call this equi-distant levels partition. In the second version we implement the t-separator using the cost reducing technique presented in Section 3. The results are shown in Table 3. We compare the cost of the separators as well as the time taken by these two versions. These tests were run on several planar graphs and typical results are shown in the table. Our experimental results show that the computation of t-separators using cost reducing technique takes slightly more time compared to computing cost separator using equi-distant levels. More importantly, the cost of the separator using the cost reducing technique is significantly smaller than that of an equi-distant separator. These results are consistent with varying the parameter t.

Table 3. Equi-distant levels vs. optimized choice of levels. We denote by *equi-cost* and *opt-cost* the cost of separators using equi-distant levels and optimized choice of levels, respectively. We denote *equi-time* and *opt-time* the computation time of Phase I+II, respectively. All times are in seconds.

Graph	$1/t = 4$	16	32	64	128	256
airfoil1($n = 4253$, $m = 11289$)						
equi-cost	125	486	744	1084	1563	2196
opt-cost	120	398	585	842	1191	1624
equi-time	0.57	0.46	0.42	0.38	0.34	0.29
opt-time	0.61	0.46	0.46	0.46	0.39	0.38
tin100000($n = 50624$, $m = 150973$)						
equi-cost	788	1910	2806	4091	5804	8245
optimized-cost	450	1344	2118	3119	4582	6447
equi-time	6.46	7.81	7.95	7.54	6.98	6.37
optimized-time	6.81	8.63	8.68	8.78	8.58	8.00

Finally, we have computed edge t-separators by assigning costs of the vertices to be their degrees and then running our vertex t-separator algorithm for graphs with weights and costs and using the cost reducing technique. Then we compute an edge separator as described in Theorem 3. Again experiments show that the size of the separators is significantly less than the bound predicted by Theorem 3. Time for computing edge separators is small, and does not vary substantially by varying the parameter t.

Table 4. Experimental results on construction of edge t-separators. Vertices have unit weights and cost reducing technique is applied. ES_t denotes the edge separator. Times are in seconds. Number of components into which the graph is partitioned and minimum and maximum weight of a component are shown.

Graph	$\frac{1}{t}$	$\frac{|ES_t|}{\sqrt{\Delta(G)/t}}$	$|ES_t|$	# of comp.	min weight	max weight	time of Ph. I + II	time of Phase III
tapir	2	0.073	20	2	487	537	0.13	0.01
$n = 1024$	4	0.231	89	4	244	263	0.12	0.01
$m = 2846$	32	0.447	472	30	22	45	0.11	0.00
	128	0.505	1066	120	1	16	0.09	0.01
airfoil1	2	0.149	80	2	1778	2475	0.71	0.04
$n = 4253$	4	0.264	224	5	529	1045	0.67	0.04
$m = 12289$	32	0.503	1064	31	111	196	0.64	0.03
	128	0.582	2350	113	15	49	0.59	0.04
triangle	2	0.384	228	2	2517	2533	0.77	0.04
$n = 5050$	4	0.374	314	4	1197	1293	0.58	0.04
$m = 14850$	32	0.598	1373	30	116	236	0.71	0.04
	128	0.610	2707	112	21	78	0.67	0.04
tin100000	2	0.235	447	2	24504	26120	9.99	0.50
$n = 50624$	4	0.379	1017	4	12559	12763	9.82	0.51
$m = 150973$	32	0.598	4545	32	1245	1794	11.99	0.52
	128	0.629	9482	126	212	486	11.79	0.53

References

1. L. Aleksandrov and H. Djidjev, *Linear Algorithms for partitioning embedded graphs of bounded genus*, SIAM J. Disc. Math., Vol.9, No.1, pp.129-150, Feb.1996.
2. W. J. Camp, S. J. Plimpton, B. A. Hendrickson, and R. W. Leland. Massively parallel methods for engineering and science problems. *Communications of the ACM*, 37(4):30–41, April 1994.
3. K. Diks, H. N. Djidjev, O. Sykora, and I. Vrto. Edge separators of planar graphs and outerplanar graphs with applications. *J. Algorithms*, 34:258–279, 1993.
4. H. N. Djidjev. Partitioning planar graphs with vertex costs: Algorithms and applications. *Algorithmica*, 28(1):51–75, 2000.
5. Hristo N. Djidjev. On the problem of partitioning planar graphs. *SIAM Journal on Algebraic and Discrete Methods*, 3:229–240, 1982.
6. John. R. Gilbert, Gary L. Miller, and Shang-Hua Teng. Geometric mesh partitioning: Implementation and experiments. *SIAM Journal on Scientific Computing*, 19(6):2091–2110, 1998.
7. B. Hendrickson and R. Leland. The Chaco user's guide — version 2.0, Sandia National Laboratories, Technical Report SAND94-2692, 1994.
8. Monika R. Henzinger, Philip Klein, Satish Rao, and Sairam Subramanian. Faster shortest-path algorithms for planar graphs. *Journal of Computer and System Sciences*, 55(1):3–23, August 1997.
9. B. W. Kernighan and S. Lin. An efficient heuristic procedure for partitioning graphs. *The Bell System Technical Journal*, pages 291–307, February 1970.
10. Richard J. Lipton and Robert E. Tarjan. A separator theorem for planar graphs. *SIAM J. Appl. Math*, 36:177–189, 1979.
11. K. Mehlhorn and S. Näher. Leda, a platform for combinatorial and geometric computing. *Communications of ACM*, 38:96–102, 1995.
12. H.S. Maini, K.G. Mehrotra, C.K. Mohan, S. Ranka, Genetic algorithms for graph partitioning and incremental graph partitioning, CRPC-TR-94504, Rice University, 1994.
13. A. Pothen, H. D. Simon, and K.-P. Liou. Partitioning sparse matrices with eigenvectors of graphs. *SIAM J. Matrix Anal. Appl.*, 11(3):430–452, July 1990.
14. K. Schloegel, G. Karypis, and V. Kumar. *Graph Partitioning for High Performance Scientific Simulations*. In J. Dongarra et al., editor, CRPC Parallel Computing Handbook. Morgan Kaufmann (in press).

Maintaining Dynamic Minimum Spanning Trees: An Experimental Study[*]

Giuseppe Cattaneo[1], Pompeo Faruolo[1], Umberto Ferraro Petrillo[1], and Giuseppe F. Italiano[2]

[1] Dipartimento di Informatica e Applicazioni, Università di Salerno, Salerno, Italy
{cattaneo,umbfer,pomfar}@dia.unisa.it
[2] Dipartimento di Informatica, Sistemi e Produzione
Università di Roma "Tor Vergata", Roma, Italy
italiano@info.uniroma2.it
http://www.info.uniroma2.it/~Italiano

Abstract. We report our findings on an extensive empirical study on several algorithms for maintaining minimum spanning trees in dynamic graphs. In particular, we have implemented and tested a variant of the polylogarithmic algorithm by Holm *et al.*, sparsification on top of Frederickson's algorithm, and compared them to other (less sophisticated) dynamic algorithms. In our experiments, we considered as test sets several random, semi-random and worst-case inputs previously considered in the literature.

1 Introduction

In this paper we consider *fully dynamic graph algorithms*, namely algorithms that maintain a certain property on a graph that is changing dynamically. Usually, dynamic changes include the insertion of a new edge, the deletion of an existing edge, or an edge cost change; the key operations are edge insertions and deletions, however, as an edge cost change can be supported by an edge deletion followed by an edge insertion. The goal of a dynamic graph algorithm is to update the underlying property efficiently in response to dynamic changes. We say that a problem is *fully dynamic* if both insertions and deletions of edges are allowed, and we say that it is *partially dynamic* if only one type of operations (i.e., either insertions or deletions, but not both) is allowed. This research area has been blossoming in the last decade, and it has produced a large body of algorithmic techniques both for undirected graphs [11,12,14,18,20,21] and for directed graphs [9,10,19,23,24]. One of the most studied dynamic graph problem is perhaps the fully dynamic maintenance of a minimum spanning tree (MST)

---

[*] This work has been partially supported by the IST Programme of the EU under contract n. IST-1999-14186 (ALCOM-FT), by the Italian Ministry of University and Scientific Research (Project "Algorithms for Large Data Sets: Science and Engineering") and by CNR, the Italian National Research Council, under contract n. 01.00690.CT26.

D. Mount and C. Stein (Eds.): ALENEX 2002, LNCS 2409, pp. 111–125, 2002.
© Springer-Verlag Berlin Heidelberg 2002

of a graph [3,13,11,14,15,20,21]. This problem is important on its own, and it finds applications to other problems as well, including many dynamic vertex and edge connectivity problems, and computing the k best spanning trees. Most of the dynamic MST algorithms proposed in the literature introduced novel and rather general dynamic graph techniques, such as the partitions and topology trees of Frederickson [14,15], the sparsification technique by Eppstein *et al.* [11] and the logarithmic decomposition by Holm *et al.* [21].

Many researchers have been complementing this wealth of theoretical results on dynamic graphs with thorough empirical studies, in the effort of bridging the gap between the design and theoretical analysis and the actual implementation, experimental tuning and practical performance evaluation of dynamic graph algorithms. In particular, Alberts *et al.* [1] implemented and tested algorithms for fully dynamic connectivity problems: the randomized algorithm of Henzinger and King [18], and sparsification [11] on top of a simple static algorithm. Amato *et al.* [5] proposed and analyzed efficient implementations of dynamic MST algorithms: the partitions and topology trees of Frederickson [14, 15], and sparsification on top of dynamic algorithms [11]. Miller *et al.* [17] proposed efficient implementations of dynamic transitive closure algorithms, while Frigioni *et al.* [16] and later Demetrescu *et al.* [8] conducted an empirical study of dynamic shortest path algorithms. Most of these implementations have been wrapped up in a software package for dynamic graph algorithms [2]. Finally, Iyer *et al.* [22] implemented and evaluated experimentally the recent fully dynamic connectivity algorithm of Holm *et al.* [21], thus greatly enhancing our knowledge on the practical performance of dynamic connectivity algorithms.

The objective of this paper is to advance our knowledge on dynamic MST algorithms by following up the recent theoretical progress of Holm *et al.* [21] with a thorough empirical study. In particular: (1) we present and experiment with efficient implementations of the dynamic MST algorithm of Holm *et al.* [21], (2) we propose new simple algorithms for dynamic MST, which are not as asymptotically efficient as [21], but nevertheless seem quite fast in practice, and (3) we compare all these new implementations with previously known algorithmic codes for dynamic MST [5], such as the partitions and topology trees of Frederickson [14,15], and sparsification on top of dynamic algorithms [11].

We found the implementations contained in [22] targeted and engineered for dynamic connectivity so that an extension of this code to dynamic MST appeared to be a difficult task. After some preliminary tests, we decided to produce a completely new implementation of the algorithm by Holm *et al.*, more oriented towards dynamic MST. With this bulk of implementations, we performed extensive tests under several variations of graph and update parameters in order to gain a deeper understanding on the experimental behavior of these algorithms. To this end, we produced a rather general framework in which the dynamic graph algorithms available in the literature can be implemented and tested. Our experiments were run both on randomly generated graphs and update sequences, and on more structured (non–random) graphs and update sequences, which tried to enforce bad update sequences on the algorithms.

2   The Algorithm by Holm *et al.*

In this section we quickly review the algorithm by Holm *et al.* for fully dynamic MST. We will start with their algorithm for handling deletions only, and then sketch on how to transform this deletion-only algorithm into a fully dynamic one. The details of the method can be found in [21].

2.1   Decremental Minimum Spanning Tree

We maintain a minimum spanning forest F over a graph G having n nodes and m edges. All the edges belonging to F will be referred as tree-edges. The main idea behind the algorithm is to partition the edges of G into different levels. Roughly speaking, whenever we delete edge (x, y), we start looking for a replacement edge at the same level as (x, y). If this search fails, we consider edges at the previous level and so on until a replacement edge is found. This strategy is effective if we could arrange the edge levels so that replacement edges can be found quickly. To achieve this task, we promote to a higher level all the edges unsuccessfully considered for a replacement.

To be more precise, we associate to each edge e of G a level $\ell(e) \leq L = \lfloor (\log n) \rfloor$. For each i, we denote by F_i the sub-forest containing all the edges having level at least i. Thus, $F = F_0 \supseteq F_1 \supseteq ... \supseteq F_L$. The following invariants are maintained throughout the sequence of updates:

1. F is a maximum (w.r.t. ℓ) spanning forest of G, that is, if (v, w) is a non-tree edge, v and w are connected in $F_{l(v,w)}$
2. The maximum number of nodes in a tree in F_i is $\lfloor n/2^i \rfloor$. Thus, the maximum relevant level is L.
3. If e is the heaviest edge on a cycle C, then e has the lowest level on C.

We briefly define the two operations Delete and Replace needed to support deletions.

Delete(e) If e is not a tree edge then it is simply deleted. If e is a tree edge, first we delete it then we have to find a replacement edge that maintains the invariants listed above and keeps a minimum spanning forest F. Since F was a minimum spanning forest before, we have that a candidate replacement edge for e cannot be at a level greater than $\ell(e)$, so we start searching for a candidate at level $\ell(e)$ by invoking operation Replace($e, \ell(e)$).

Replace($(v, w), i$) Assuming there is no replacement edge on level $> i$, finds a replacement edge of the highest level $\leq i$, if any. The algorithm works as follows. Let T_v and T_w be the tree in F_i containing v and w, respectively. After deleting edge (v, w), we have to find the minimum cost edge connecting again T_v and T_w. First, we move all edges of level i of T_v to level $(i+1)$, then, we start considering all edges on level i incident to T_v in a non-decreasing weight order. Let f be the edge currently considered: if f does not connect T_v and T_w, then we promote f to level $(i + 1)$ and we continue the search. If f connects T_v and T_w, then it is inserted as a replacement edge and the search stops. If the search fails, we call Replace($(u, w), i - 1$). When the search fails also at level 0, we stop as there is no replacement edge for (v, w).

2.2 The Fully Dynamic Algorithm

Starting from the deletions-only algorithm, Holm *et al.* obtained a fully dynamic algorithm by using a clever refinement of a technique by Henzinger and King [20] for developing fully dynamic data structures starting from deletions-only data structures.

We maintain the set of data structures $\mathcal{A} = A_1, .., A_{s-1}, A_s$, $s = \lceil (\log n) \rceil$, where each A_i is a subgraph of G. We denote by F_i the local spanning forest maintained on each A_i. We will refer to edges in F_i as *local tree edges* while we will refer to edges in F as *global tree edges*. All edges in G will be in at least one A_i, so we have $F \subseteq \bigcup_i F_i$. During the algorithm we maintain the following invariant:

1. For each global non-tree edge $f \in G \backslash F$, there is exactly one i such that $f \in A_i \backslash F_i$ and if $f \in F_j$, then $j > i$.

Without loss of generality assume that the graph is connected, so that we will talk about MST rather than minimum spanning forest. At this point, we use a dynamic tree of Sleator and Tarjan [28] to maintain the global MST and to check if update operations will change the solution. Here is a brief explanation of the update procedures:

Insert(e) Let be $e = (v, w)$, if v and w are not connected in F by any edge then we add e to F. Otherwise, we compare the weight of e with the heaviest edge on the path from v to w. If e is heavier, we just update $\mathcal{A}$ with e, otherwise we replace f with e in F and we call the update procedure on $\mathcal{A}$.

Delete(e) We delete e from all the A_i and we collect in a set R all the replacement edges returned from each deletions-only data structure. Then, we check if e is F, if so we search in R the minimum cost edge reconnecting F. Finally, we update $\mathcal{A}$ using R.

Update $\mathcal{A}$ with edge set D We find the smallest j such that $|(D \bigcup_{h \leq j}(A_h \backslash F_h)) \backslash F| \leq 2^j$. Then we set

$$A_j = F \cup D \cup \bigcup_{h \leq j}(A_h \backslash F_h)),$$

and we initialize A_j as a MST deletions-only data structure. Finally we set $A_h = 0$ for all $h < j$.

The initialization required by the updates is one of the crucial points in the algorithm by Holm *et al.* To perform this task efficiently, they use a particular compression of some subpaths, which allows one to bound the initialization work at each update. This compression is carried out with the help of top trees: details can be found in [21].

2.3 Our Implementation

Our implementation follows exactly the specifications of Holm *et al.* except for the use of the compression technique described in [21]. Indeed, our first experience with the compression and top trees was rather discouraging from the experimental viewpoint. In particular, the memory requirement was substantial so that we could not experiment with medium to large size graphs (order of thousands of vertices). We thus engineered a different implementation of the compression technique using the dynamic trees of Sleator and Tarjan [28] in place of the top trees; this yielded a consistent gain on the memory requirements of the resulting algorithm with respect to our original implementation. We are currently working on a more sophisticated implementation of top trees that should allow us to implement in a faster way the original compression.

3   Simple Algorithms

In this section we describe two simple algorithms for dynamic MST that are used in our experiments. They are basically a fast "dynamization" of the static algorithm by Kruskal (see e.g., [7,25]). We recall here that Kruskal's algorithm grows a forest by scanning all the graph edges by increasing cost: if an edge (u, v) joins two different trees in the forest, (u, v) is kept and the two trees are joined. Otherwise, u and v are already in a same tree, and (u, v) is discarded.

3.1   ST-Based Dynamic Algorithm

Our dynamization of the algorithm by Kruskal uses the following ideas. Throughout the sequence of updates, we keep the following data structures: the minimum spanning tree is maintained with a dynamic tree of Sleator and Tarjan [28], say MST, while non-tree edges are maintained sorted in a binary search tree, say NT.

When a new edge (x, y) is to be inserted, we check in $O(\log n)$ time with the help of dynamic trees whether (x, y) will become part of the solution. If this is the case, we insert (x, y) into MST: the swapped edge will be deleted from MST and inserted into NT. Otherwise, (x, y) will not be a tree edge and we simply insert it into NT.

When edge (x, y) has to be deleted, we distinguish two cases: if (x, y) is a non-tree edge, then we simply delete it from NT in $O(\log n)$. If (x, y) is a tree edge, its deletion disconnects the minimum spanning tree into T_x (containing x) and T_y (containing y), and we have to look for a replacement edge for (x, y). We examine non-tree edges in NT by increasing costs and try to apply the scanning strategy of Kruskal on T_x and T_y: namely, for each non-tree edge $e = (u, v)$, in increasing order, we check whether e reconnects T_x and T_y: this can be done via findroot(u) and findroot(v) operations in Sleator and Tarjan's trees. Whenever such an edge is found, we insert it into MST and stop our scanning. The total time required by a deletion is $O(k \cdot \log n)$, where k is the total number of non-tree edges scanned. In the worst case, this is $O(m \log n)$.

We refer to this algorithm as ST. Note that ST requires few lines of code, and fast data structures, such as the dynamic trees [28]. We therefore expect it to be very fast in practice, especially in update sequences containing few tree edge deletions or in cases when, after deleting tree edge (x, y), the two trees T_x and T_y get easily reconnected (e.g., the cut defined by (x, y) contains edges with relatively small costs).

3.2   ET-Based Dynamic Algorithm

Our first experiments with ST showed that it was indeed fast in many situations. However, the most difficult cases for ST were on sparse graphs, i.e., graphs with around n vertices. In particular, the theory of random graphs [6] tells us that when $m = n/2$ the graph consists of a few large components of size $O(n^{2/3})$ and smaller components of size $O(\log n)$. For $m = 2n$, the graph contains a giant component of size $O(n)$ and smaller components of size $O(\log n)$. In these random cases, a random edge deletion is likely to disconnect the minimum spanning forest and to cause the scanning of many non-tree edges in the quest for a replacement. Indeed, a careful profiling showed that most of the time in these cases was spent by ST in executing findroot operations on Sleator and Tarjan's trees (ST trees).

We thus designed another variant of this algorithm, referred to as ET, which uses Euler Tour trees (ET trees) in addition to ST trees. The ET-tree (see e.g., [18] for more details) is a balanced search tree used to efficiently maintain the Euler Tour of a given tree. ET-trees have some interesting properties that are very useful in dynamic graph algorithms. In our implementation, we used the randomized search tree of Aragon and Seidel [4] to support the ET-trees.

In particular, we keep information about tree edges both with an ST tree and with an ET tree. The only place were we use the ET-tree is in the deletion algorithm, i.e., where we check whether a non-tree edge $e = (u, v)$ reconnects the minimum spanning tree. We still need ST-trees, however, as they give a fast method for handling edge insertions. Note that ET has the same asymptotical update bounds as ST.

The main difference between the two implementations is that we expect findroot operations on randomized search trees to be faster than on Sleator and Tarjan's trees, and consequently ET to be faster than ST on sparse graphs. However, when findroot operations are no longer the bottleneck, we also expect that the overhead of maintaining both ET-trees and ST-trees may become significant. This was exactly confirmed by our experiments, as shown in Fig. 3.2, which illustrates an experiment on random graphs with 2,000 vertices and different densities.

3.3   Algorithms Tested

We have developed, tested and engineered many variants of our implementations. All the codes are written in C++ with the support of the LEDA [26] algorithmic software library, and are homogeneously written by the same people, i.e., with the same algorithmic and programming skills. In this extended abstract, we will report only on the following four implementations:

HDT   Our implementation of the algorithm proposed by Holm *et al.* with the original deletions-only fully-dynamic transformation proposed by Henzinger and King. It uses randomized balanced ET Trees from [1].

Spars   Simple sparsification run on top of Frederickson's light partition of order $\lceil m^{2/3} \rceil$ and on lazy updates, as described in [5].

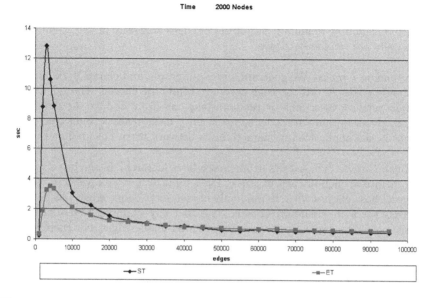

Fig. 1. ET and ST on random graphs with 2,000 vertices and different densities. Update sequences were random and contained 10,000 edge insertions and 10,000 edge deletions.

ST   The implementation of the algorithm described in Sect. 3.1.
ET   The implementation of the algorithm described in Sect. 3.2.

4   Experimental Settings

We have conducted our tests on a Pentium III (650 MHz) under Linux 2.2 with 768 MB of physical RAM, 16 KB L1-cache and 256 KB L2-cache. We choose this platform as it was the one with largest RAM available to us, so that we could experiment with large graphs without incurring in external memory swap problems. Our codes were compiled with a Gnu C++ 2.9.2 compiler with the -O flag. The metrics we have used to rate the algorithms' performance were the total amount of CPU time in user mode spent by each process and the maximum amount of memory allocated during the algorithms' execution.

In all our experiments, we generated a weighted graph $G = (V, E)$, and a sequence σ of update operations featuring i edge insertions and d edge deletions on this input graph G. We then fed each algorithm to be tested with G and the sequence σ. All the collected data were averaged on ten different experiments. Building on previous experimental work on dynamic graph algorithms [1,5,22], we considered the following test sets:

Random Graphs. The initial graph G is generated randomly according to a pair of input parameters (n, m), where n is the initial number of nodes and m the initial number of edges of G. To generate the update sequence, we choose at random an edge to insert from the edges not currently in the graph or an edge to delete, again at random from the set of edges currently in the graph. All the edge costs are randomly chosen.

Semirandom Graphs. We generate a random graph, and choose a fix number of candidate edges E. All the edge costs are randomly chosen. The updates here contain random insertions or random deletions from E only. As pointed out in [22], this semirandom model seems slightly more realistic than true random graphs in the application of maintaining a network when links fail and recover.

k-Clique Graphs. These tests define a two-level hierarchy in the input graph G: we generate k cliques, each of size c, for a total of $n = k \cdot c$ vertices. We next connect those cliques with $2k$ randomly chosen inter-clique edges. As before, all the edge costs are randomly chosen. Note that any spanning tree of this graph consists of intra-clique trees connected by inter-clique edges. For these k-clique graphs, we considered different types of updates. On the one side, we considered operations involving inter-clique edges only (i.e., deleting and inserting inter-clique tree edges). Since the set of replacement edges for an inter-clique tree edge is very small, this sequence seems particularly challenging for dynamic MST algorithms. The second type of update operations involved the set of edges inside a clique (intra-clique) as well and considered several kinds of mix between inter-clique and intra-clique updates.

As reported in [22], this family of graphs seems interesting for many reasons. First, it has a natural hierarchical structure, common in many applications, and not found in random graphs. Furthermore, investigating several combinations of inter-clique/intra-clique updates on the same k-clique graph can stress algorithms on different terrains, ranging from worst-case inputs (inter-clique updates only) to more mixed inputs (both inter- and intra-clique updates).

Worst-case inputs. For sake of completeness, we adapted to minimum spanning trees also the worst-case inputs introduced by Iyer *at al.* [22] for connectivity. These are inputs that try to force a bad sequence of updates for HDT, and in particular try to promote as many edges as possible through the levels of its data structures. We refer the interested reader to reference [22] for the full details. For the sake of completeness, we only mention here that in these test sets there are no non-tree edges, and only tree edge deletions are supported: throughout these updates, HDT is foiled by unneeded tree edge movements among levels.

5   Experimental Results

Random inputs. Not surprisingly, on random graphs and random updates, the fastest algorithms were ST and ET. This can be easily explained, as in this case edge updates are not very likely to change the MST, especially on dense graphs: ST and ET are very simple algorithms, can be coded in few lines, and therefore likely to be superior in such a simple scenario. The only interesting issue was perhaps the case of sparse random graphs, where a large number of updates can force changes in the solution. As already explained in Sect. 3.2, ST does not perform well in such a scenario, while ET exhibits a more stable behavior. As already observed in [22], the decomposition into edge levels of [21] helps HDT "learn" about and adapt to the structure of the underlying graph. A random graph has obviously no particular structure, and thus all the machinery of HDT would not seem particularly helpful in this case. It was interesting to notice, however, that even in this experiment, HDT was slower than ET and ST only by a factor of 5. As far as Spars is concerned, it was particularly appealing for sparse graphs, when the sparsification tree had a very small height. As the number of edges increased, the overhead of the sparsification tree became more significant. Figure 2 illustrates the results of our experiments with random graphs with 2,000 vertices and different densities. Other graph sizes exhibited a similar behavior.

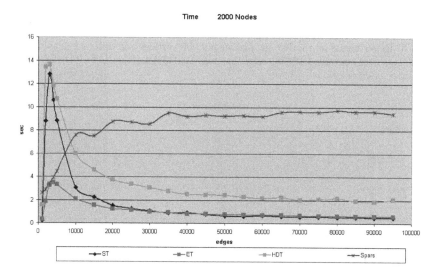

Fig. 2. Experiments on random graphs with 2,000 vertices and different densities. Update sequences contained 10,000 insertions and 10,000 deletions.

We also experimented with the operations by changing the sequence mix: to force more work to the algorithms, we increased the percentage of tree edge deletions. Indeed, this seems to be the hardest operation to support, as non-tree deletion is trivial, and tree insertions can be easily handled with ST trees. On random graphs, ET remained consistently stable even with evenly mixed update sequences (50% insertions, 50% deletions) containing 45% of tree edge deletions. ST was penalized more in this, because of the higher cost of findroot on ST-trees. Figure 3 reports the results of such an experiment.

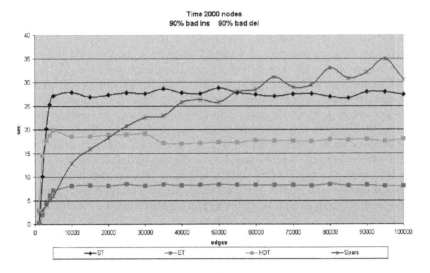

Fig. 3. Experiments on random graphs with 2,000 vertices and different densities. Update sequences contained 10,000 insertions and 10,000 deletions: 45% of the operations were tree edge deletions.

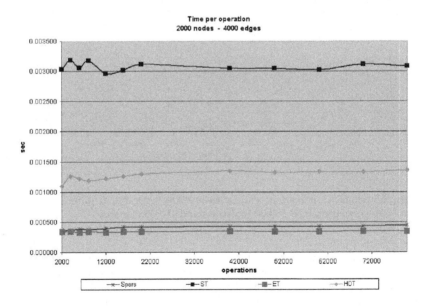

Fig. 4. Experiments on semirandom graphs with 2,000 vertices and 4,000 edges. The number of operations ranges from 2,000 to 80,000.

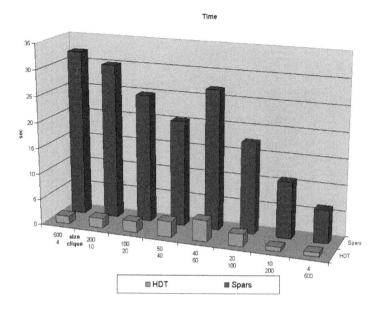

Fig. 5. Experiments on k-cliques graphs with inter-clique operations only.

Semirandom inputs. We have made several experiments with semirandom inputs using random graphs with 2,000 vertices and a number of operations ranging from 2,000 to 8,000. In the first experiment we have chosen a subset E of 1,000 edges. As already pointed out by Holm *et al* in [22], in this case we have many disconnected components that will never be joined again since the edge set E is fixed. The performance of ST and ET seems quite good since very few non-tree edges are candidates for being replacement edges; the same behavior seems to apply to HDT. On the other hand, here Spars performs very badly compared to the other algorithms: although the graphs we are experimenting with are very sparse, in this case there is a high overhead due to the underlying partition of Frederickson. This situation totally changes when we increase the size of E. In our second and third experiment we have fixed the E size respectively to 2,000 and 4,000 edges. According to our results illustrated in Fig. 4, ET still remained the fastest algorithm together with Spars that seemed to be the most stable algorithm among the one we tested for this kind of experiments. ST and HDT suffered a significant performance loss. In the case of ST, its behavior has been very similar to the one we measured in the Random experiments and is probably due to the overhead of findroots.

k-Clique inputs. For these tests, we considered two kinds of update sequences, depending on whether all update sequences were related only to inter-clique edges or could be related to intra-clique edges as well. In the first case, where only inter-clique edges were involved HDT was by far the quickest implementation. In

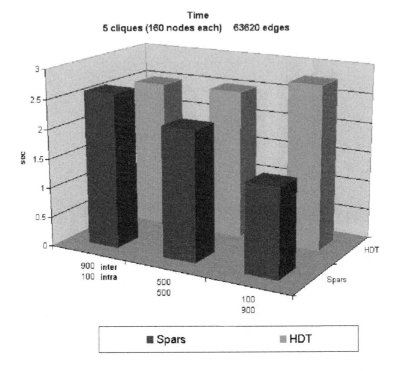

Fig. 6. Experiments on k-cliques graphs with a different mix of inter- and intra-clique operations.

fact, stimulated by inter-clique updates, HDT was quite fast in learning the 2-level structure of the graph, and in organizing its level decomposition accordingly, as shown in Fig. 5.

When updates could involve also intra-clique edges as well, however, the random structure of the update sequence was somehow capable of hiding from HDT the graph structure, thus hitting its level decomposition strategy. Indeed, as it can be seen from Fig. 6, as the number of operations involving intra-clique edges increased, the performance of Spars improved (updates on intra-clique edges are not likely to change the MST and thus will not propagate all the way up to the sparsification tree root), while on the contrary the performance of HDT slightly deteriorated.

In both cases, ET and ST were not competitive on these test sets, as the deletion of an inter-clique edge, for which the set of replacement edges is very small, could have a disastrous impact on those algorithms.

Worst-case inputs. Figure 7 illustrates the results of these tests on graphs with up to 32,768 vertices. As expected, HDT is tricked by the update sequence and spends a lot of time in (unnecessary!) promotions of edges among levels of the

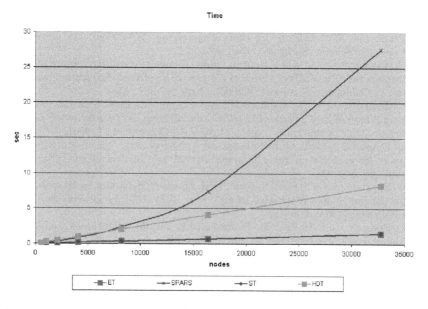

Fig. 7. Experiments on worst-case inputs on graphs with different number of vertices.

data structures. ET and ST achieve their best case of $O(\log n)$ time, as there are no non-tree edges to consider. Spars is also hit pretty badly by these test sets: indeed each tree edge deletion suffers from the overhead of the underlying implementation of Frederickson's light partition of order $\lceil m^{2/3} \rceil$.

References

1. D. Alberts, G. Cattaneo, G. F. Italiano, "An empirical study of dynamic graph algorithms", *ACM Journal on Experimental Algorithmics*, vol. 2 (1997).
2. D. Alberts, G. Cattaneo, G. F. Italiano, U. Nanni, C. D. Zaroliagis "A Software Library of Dynamic Graph Algorithms". *Proc. Algorithms and Experiments* (ALEX 98), Trento, Italy, February 9–11, 1998, R. Battiti and A. A. Bertossi (Eds), pp. 129–136.
3. D. Alberts, M. R. Henzinger, "Average Case Analysis of Dynamic Graph Algorithms", *Proc. 6th Symp. on Discrete Algorithms* (1995), 312–321.
4. C.R. Aragon, R. Seidel, "Randomized search trees", *Proc. 30th Annual Symp. on Foundations of Computer Science* (FOCS 89), 1989, pp. 540–545.
5. G. Amato, G. Cattaneo, G. F. Italiano, "Experimental Analysis of Dynamic Minimum Spanning Tree Algorithms", *Proc. 8th Annual ACM-SIAM Symposium on Discrete Algorithms*, (1997), 5–7.
6. B. Bollobás. *Random Graphs*. Academic Press, London, 1985.
7. T. H. Cormen, C. E. Leiserson, R. L. Rivest. *Introduction to Algorithms*. Mac-Graw Hill, NY, 1990.

8. C. Demetrescu, D. Frigioni, A. Marchetti-Spaccamela, U. Nanni. "Maintaining Shortest Paths in Digraphs with Arbitrary Arc Weights: An Experimental Study." *Procs. 3rd Workshop on Algorithm Engineering* (WAE2000). Lecture Notes in Computer Science, Springer-Verlag, 2001.

9. C. Demetrescu and G.F. Italiano. Fully dynamic transitive closure: Breaking through the O(n2) barrier. *Proc. of the 41st IEEE Annual Symposium on Foundations of Computer Science* (FOCS'00), pages 381-389, 2000.

10. C. Demetrescu, G. F. Italiano, "Fully Dynamic All Pairs Shortest Paths with Real Edge Weights". *Proc. 42nd IEEE Annual Symp. on Foundations of Computer Science* (FOCS 2001), Las Vegas, NV, U.S.A., October 14-17, 2001.

11. D. Eppstein, Z. Galil, G. F. Italiano and A. Nissenzweig, *"Sparsification – A technique for speeding up dynamic graph algorithms"*. In *Journal of ACM*, Vol. 44, 1997, pp. 669-696.

12. D. Eppstein, Z. Galil, G. F. Italiano, T. H. Spencer, "Separator based sparsification for dynamic planar graph algorithms", *Proc. 25th ACM Symposium on Theory of Computing* (1993), 208–217.

13. D. Eppstein, G. F. Italiano, R. Tamassia, R. E. Tarjan, J. Westbrook, and M. Yung, Maintenance of a minimum spanning forest in a dynamic plane graph, *J. Algorithms*, 13:33–54, 1992.

14. G.N. Frederickson, "Data structures for on-line updating of minimum spanning trees, with applications", *SIAM J. Comput.* 14 (1985), 781–798.

15. G.N. Frederickson, "Ambivalent data structures for dynamic 2-edge-connectivity and k smallest spanning trees", *Proc. 32nd IEEE Symp. Foundations of Computer Science* (1991), 632–641.

16. D. Frigioni, M. Ioffreda, U. Nanni, G. Pasqualone. "Experimental Analysis of Dynamic Algorithms for the Single Source Shortest Path Problem." *ACM Journal on Experimental Algorithmics*, vol. 3 (1998), Article 5.

17. D. Frigioni, T. Miller, U. Nanni, G. Pasqualone, G. Schaefer, and C. Zaroliagis "An experimental study of dynamic algorithms for directed graphs", in *Proc. 6th European Symp. on Algorithms*, Lecture Notes in Computer Science 1461 (Springer-Verlag, 1998), pp.368-380.

18. M. R. Henzinger and V. King, Randomized dynamic graph algorithms with poly-logarithmic time per operation,*Proc. 27th Symp. on Theory of Computing*, 1995, 519–527.

19. M. R. Henzinger and V. King, Fully dynamic biconnectivity and transitive closure, *Proc. 36th IEEE Symp. Foundations of Computer Science*, pages 664–672, 1995.

20. M. R. Henzinger and V. King, Maintainig minimum spanning trees in dynamic graphs, *Proc. 24th Int. Coll. Automata, Languages and Programming* (ICALP 97), pages 594–604, 1997.

21. J. Holm, K. de Lichtenberg, and M. Thorup, Poly-logarithmic deterministic fully-dynamic algorithms for connectivity, minimum spanning tree, 2-edge, and biconnectivity. *Proc. 30th Symp. on Theory of Computing* (1998), pp. 79-89.

22. R. Iyer, D. R. Karger, H. S. Rahul, and M. Thorup, An Experimental Study of Poly-Logarithmic Fully-Dynamic Connectivity Algorithms, *Proc. Workshop on Algorithm Engineering and Experimentation*, 2000.

23. V. King. Fully dynamic algorithms for maintaining all-pairs shortest paths and transitive closure in digraphs. *Proc. 40th IEEE Symposium on Foundations of Computer Science* (FOCS'99), 1999.

24. V. King and G. Sagert. A fully dynamic algorithm for maintaining the transitive closure. *Proc. 31st ACM Symposium on Theory of Computing* (STOC'99), pages 492-498, 1999.

25. J. B. Kruskal, On the shortest spanning subtree of a graph and the traveling salesman problem, *Proc. Amer. Math. Soc. 7*, (1956), 48–50.

26. K. Melhorn and S. Näher, *"LEDA, A platform for combinatorial and geometric computing"*. *Comm. ACM*, (1995), 38(1): pp.96-102.

27. M. Rauch, "Improved data structures for fully dynamic biconnectivity", *Proc. 26th Symp. on Theory of Computing* (1994), 686–695.

28. D. D. Sleator and R. E. Tarjan, A data structure for dynamic trees, *J. Comp. Syst. Sci.*, 24:362–381, 1983.

29. R. E. Tarjan and J. van Leeuwen, Worst-case analysis of set union algorithms, *J. Assoc. Comput. Mach.*, 31:245–281, 1984.

Experimental Evaluation of a New Shortest Path Algorithm*
(Extended Abstract)

Seth Pettie, Vijaya Ramachandran, and Srinath Sridhar

Department of Computer Sciences
The University of Texas at Austin
Austin, TX 78712
seth@cs.utexas.edu, vlr@cs.utexas.edu, srinath@cs.utexas.edu

Abstract. We evaluate the practical efficiency of a new shortest path algorithm for undirected graphs which was developed by the first two authors. This algorithm works on the fundamental *comparison-addition model*.
Theoretically, this new algorithm out-performs Dijkstra's algorithm on sparse graphs for the all-pairs shortest path problem, and more generally, for the problem of computing single-source shortest paths from $\omega(1)$ different sources. Our extensive experimental analysis demonstrates that this is also the case in practice. We present results which show the new algorithm to run faster than Dijkstra's on a variety of sparse graphs when the number of vertices ranges from a few thousand to a few million, and when computing single-source shortest paths from as few as three different sources.

1   Introduction

The shortest paths problem on graphs is one of the most widely-studied combinatorial optimization problems. Given an edge-weighted graph, a path from a vertex u to a vertex v is a *shortest path* if its total length is minimum among all u-to-v paths. The complexity of finding shortest paths seems to depend upon how the problem is formulated and what kinds of assumptions we place on the graph, its edge-lengths and the machine model. Most shortest path algorithms for graphs can be well-categorized by the following choices.

1. Whether shortest paths are computed from a single *source* vertex to all other vertices (SSSP), or between all pairs of vertices (APSP). One should also consider the intermediate problem of computing shortest paths from *multiple* specified sources (MSSP).
2. Whether the edge lengths are non-negative or arbitrary.

---

* This work was supported by Texas Advanced Research Program Grant 003658-0029-1999 and NSF Grant CCR-9988160. Seth Pettie was also supported by an MCD Graduate Fellowship.

D. Mount and C. Stein (Eds.): ALENEX 2002, LNCS 2409, pp. 126–142, 2002.
© Springer-Verlag Berlin Heidelberg 2002

3. Whether the graph is directed or undirected.

4. Whether shortest paths are computed using just *comparison & addition* operations, or whether they are computed assuming a specific edge-length representation (typically integers in binary) and operations specific to that representation. Comparison-addition based algorithms are necessarily general and they work when edge-lengths are either integers or real numbers.

There is a wealth of literature on variations of the shortest path problem,[1] however despite such intense research, very few of the results beyond the classical algorithms of Dijkstra, Bellman-Ford, Floyd-Warshall, and min-plus matrix multiplication [AHU74,CLR90] work with real-valued edge-lengths using only comparisons and additions.[2]

Previous experimental studies of shortest path algorithms [CGR96,GS97, G01b] focussed on very restricted classes of inputs, where the edge lengths were assumed to be uniformly distributed, relatively small integers. This approach may be preferable for a specific application, however any algorithm implemented for more general use must be *robust*. By robust we mean that it makes no assumptions on the distribution of inputs, and minimal assumptions on the *programming interface* to the input (in the case of shortest path problems this leads naturally to the comparison-addition model); we elaborate on this in Section 2. A fact which many find startling is that Dijkstra's 1959 algorithm is still the best robust SSSP & APSP algorithm for positively-weighted sparse directed graphs.

In this paper we evaluate the performance of the recent *undirected* shortest path algorithm of Pettie & Ramachandran [PR02], henceforth the *PR algorithm*. The PR algorithm is a robust, comparison-addition based algorithm for solving undirected SSSP from multiple specified sources (MSSP). It works by pre-computing a certain structure called the 'component hierarchy', or *CH* (first proposed by Thorup [Tho99], for use with integer edge lengths) in time $O(m + n \log n)$. Once the CH is constructed SSSP is solved from any source in $O(m\alpha(m, n))$ time, where α is the very slow-growing inverse-Ackermann function. Theoretically this algorithm is asymptotically faster than Dijkstra's when the number of sources is $\omega(1)$ and the number of edges is $o(n \log n)$.

The PR algorithm (as well as [Tho99,Hag00]) can also tolerate a dynamic graph in some circumstances. If a component hierarchy is constructed for a graph G, SSSP can be solved in $O(m\alpha(m, n))$ time on any graph G' derived from G by altering each edge weight by up to a constant factor.

As mentioned above, there are only a few shortest path algorithms that work on the comparison-addition model, and there is only one robust algorithm in direct competition with PR, namely Dijkstra's. The Bellman-Ford, Floyd-Warshall, and min-plus matrix multiplication algorithms handle negative edge

---

[1] For an up-to-date survey of shortest path algorithms, see Zwick [Z01] (an updated version is available on-line).

[2] Some exceptions to this rule are Fredman's min-plus matrix multiplication algorithm [F76] and several algorithms with good *average-case* performance: [MT87,KKP93, KS98,Mey01,G01]

lengths and as a consequence are considerably less efficient than the PR algorithm (quadratic time for SSSP and cubic for APSP). The fastest implementation [Tak92] of Fredman's algorithm [F76] for APSP also takes almost cubic time. The average-case algorithms in [KKP93,McG91,Jak91,MT87] only provide improvements on very dense random graphs.

We evaluate the practical efficiency of the PR algorithm for the MSSP problem on undirected graphs by comparing it with Dijkstra's algorithm. The MSSP problem generalizes the SSSP-APSP extremes, and could be more relevant in some practical scenarios. For instance, a recent algorithm of Thorup [Tho01] for the graphic facility location and k-median problems performs SSSP computations from a polylog number of sources. Our experiments indicate quite convincingly that the Pettie-Ramachandran algorithm outperforms Dijkstra on sparse graphs when computing SSSP from a sufficient number of sources, as few as 3 or 4 in several cases. We obtained this result across all classes of sparse graphs that we considered except for the so-called 'long grids' [CGR96]. We also compare the PR algorithm to breadth first search, a natural lower bound on SSSP and a useful routine to normalize the running times of shortest path algorithms across different architectures. We elaborate on this and other aspects of our results in Section 6. Clearly, our results also apply to the APSP problem, and they show that the PR algorithm outperforms Dijkstra's algorithm for the APSP problem on sparse graphs.

The rest of the paper is organized as follows. In Section 2 we delineate the scope of our study. In Section 3 we give an overview of Dijkstra's algorithm and the PR algorithm. Section 4 describes the design choices we made in implementing the two algorithms. Section 5 describes our experimental set-up, and Section 5.1 the types of graphs we used. Section 6 provides our results. Section 7 ends with a discussion.

2   Scope of This Work

The focus of this paper is *robust* shortest path algorithms, so it is worthwhile to state here exactly what we mean by the term. A robust shortest path algorithm should be robust with respect to:

Input format. The algorithm should work with minimal assumptions on the input format and the programming "hooks" to manipulate the input. The assumption that edge-lengths are subject to comparison and addition operations is minimal since these operations are both necessary and sufficient to solve shortest path problem.

Graph type. The algorithm should work well on *all* graph sizes & topologies. It should not depend on the graph being overly structured (e.g. grids) or overly random (e.g. the $G_{n,m}$ distr.).

Edge-length distribution. The algorithm should not be adversely affected by the range or distribution on edge-lengths, nor should it depend upon the edge-lengths being chosen independently at random.

Some may object to the first criterion because, at some level, edge lengths are represented as `ints` or `doubles`; one might as well assume such an input. This is not quite true. For instance, the LEDA platform [MN99] uses different types for rationals, high-precision floating point numbers, and 'real' numbers with provable accuracy guarantees, and Java has similar types BigDecimal and BigInteger. A robust algorithm can be used with all such types with little or no modification, and can be ported to different platforms with minimal modifications.

The bottom line is that robust algorithms are fit for use in a general setting where the format and distribution of inputs is unknown and/or varies. Nothing precludes the use of other specialized shortest path algorithms (indeed, those tailored to small integer weights, e.g. [GS97], will likely be faster), however, depending solely on such an algorithm is clearly unwise.

In our experiments we focus primarily on classes of *sparse graphs*, which we define as having an edge-to-vertex ratio less than $\log n$. Sparse graphs frequently arise naturally; e.g. all planar and grid-like graphs are sparse, and the evidence shows the 'web graph' also to be sparse. Denser graphs are important as well, but as a practical matter the SSSP problem has essentially been solved: Dijkstra's algorithm runs in linear time for densities greater than $\log n$. The "sorting bottleneck" in Dijkstra's algorithm is only apparent for sparse graphs.

3   Overview of the Algorithms

Dijkstra's algorithm [Dij59] for SSSP (see [CLR90] or [AHU74]) visits the vertices in order of increasing distance from the source. It maintains a set S of visited vertices whose distance from the source has been established, and a tentative distance $D(v)$ to each unvisited vertex v. $D(v)$ is an upper bound on the actual distance to v, denoted $d(v)$; it is the length of the shortest path from the source to v in the subgraph induced by $S \cup \{v\}$. Dijkstra's algorithm repeatedly finds the unvisited vertex with minimum tentative distance, adds it to the set S and updates D-values appropriately.

Rather than giving a description of the Pettie-Ramachandran [PR02] algorithm (which is somewhat involved), we will instead describe the component hierarchy *approach* put forward by Thorup [Tho99]. Suppose that we are charged with finding all vertices within distance b of the source, that is, all v such that $d(v) \in [0, b)$. One method is to run Dijkstra's algorithm (which visits vertices in order of their d-value) until a vertex with d-value outside $[0, b)$ is visited. Thorup observed that if we choose $t < b$ and find the graph G_t consisting of edges shorter than t, the connected components of G_t, say $\mathcal{G}_t$, can be dealt with separately in the following sense. We can simulate which vertices Dijkstra's algorithm *would* visit for each connected component in $\mathcal{G}_t$, first over the interval $[0, t)$, then $[t, 2t)$, $[2t, 3t)$, up to $[\lfloor \frac{b}{t} \rfloor t, b)$. It is shown in [Tho99] (see also [PR02]) that these separate subproblems do not "interfere" with each other in a technical sense. The subproblems generated by Thorup's approach are solved recursively. The component hierarchy is a rooted tree which represents how the graph is decomposed; it is determined by the underlying graph and choices of t made in the algorithm.

The basic procedure in component hierarchy-based algorithms [Tho99,Hag00, PR02] is Visit(x, I), which takes a component hierarchy node x and an interval I, and visits all vertices in the subgraph corresponding to x whose d-values lie in I.

4 Design Choices

4.1 Dijkstra's Algorithm

We use a *pairing heap* [F+86] to implement the priority queue in Dijkstra's algorithm. We made this choice based on the results reported in [MS94] for minimum spanning tree (MST) algorithms. In that experiment the pairing heap was found to be superior to the Fibonacci heap (the choice for the theoretical bound), as well as d-ary heaps, relaxed heaps and splay heaps in implementations of the Prim-Dijkstra MST algorithm.[3] Since the Prim-Dijkstra MST algorithm has the same structure as Dijkstra's SSSP algorithm (Dijkstra presents both of these algorithms together in his classic paper [Dij59]), the pairing heap appears to be the right choice for this algorithm.

The experimental studies by Goldberg [CGR96,GS97,G01b] have used buckets to implement the heap in Dijkstra's algorithm. However, the bucketing strategy they used applies only to integer weights. The bucketing strategies in [Mey01, G01] could apply to arbitrary real edge weights, but they are specifically geared to good performance on edge-weights uniformly distributed in some interval. The method in [G01] can be shown to have bad performance on some natural inputs.[4] In contrast we are evaluating robust, general-purpose algorithms that function in the comparison-addition model.

We experimented with two versions of Dijkstra's algorithm, one which places all vertices on the heap initially with key value ∞ (the traditional method), and the other that keeps on the heap only vertices known to be at finite distance from the source. For sparse graphs one would expect the heap to contain fewer vertices if the second method is used, resulting in a better running time. This is validated by our experimental data. The second method out-performed the first one in all graphs that we tested, so we report results only for the second method.

4.2 Pettie-Ramachandran Algorithm

The primary consideration in [PR02] was asymptotic running time. In our implementation of this algorithm we make several simplifications and adjustments which are more practical but may deteriorate the worst-case asymptotic performance of the algorithm.

---

[3] This algorithm was actually discovered much earlier by Jarník [Jar30].

[4] For instance, where each edge length is chosen independently from one of two uniform distributions with very different ranges.

1. Finding MST:

   The [PR02] algorithm either assumes the MST is found in $O(m + n \log n)$ time (for the multi-source case) or, for the single source case, in optimal time using the algorithm of [PR00]. Since, for multiple sources, we both find and sort the MST edges, we chose to use Kruskal's MST algorithm, which runs in $O(m \log n)$ time but does both of these tasks in one pass. Some of our data on larger and denser graphs suggests that it may be better to use the Prim-Dijkstra MST algorithm, which is empirically faster than Kruskal's [MS94], followed by a step to sort only the MST edges.

2. Updating D-values:

   In [PR02] the D-value of an internal CH node is defined to be the minimum D-value over its descendant leaves. As leaf D-values change, the internal D-values must be updated. Rather than use Gabow's near-linear time data structure [G85], which is rather complicated, we use the naïve method. Whenever a leaf's D-value decreases, the new D-value is propagated up the CH until an ancestor is reached with an even lower D-value. The worst-case time for updating a D-value is clearly the height of CH, which is $\log R$, where R is the ratio of the maximum to minimum edge-weight; on the other hand, very few ancestors need to be updated in practice.

3. Using Dijkstra on small subproblems:

   The stream-lined nature of Dijkstra's algorithm makes it the preferred choice for computing shortest paths on small graphs. For this reason we revert to Dijkstra's algorithm when the problem size becomes sufficiently small. If $\texttt{Visit}(x, I)$ is called on a CH node x with fewer than ν descendant leaves, we run Dijkstra's algorithm over the interval I rather than calling $\texttt{Visit}$ recursively. For *all* the experiments described later, we set $\nu = 50$.

4. Heaps vs. Lazy Bucketing:

   The [PR02] algorithm implements a priority queue with a *comparison-addition based* 'lazy bucketing' structure. This structure provides asymptotic guarantees, but for practical efficiency we decided to use a standard pairing heap to implement the priority queue, augmented with an operation called *threshold* which simulates emptying a bucket. A call to *threshold(t)* returns a list of all heap elements with keys less than t. It is implemented with a simple DFS of the pairing heap. An upper bound on the time for threshold to return k elements is $O(k \log n)$, though in practice it is much faster.

5. Additional Processing of CH:

   In [PR02, Sections 3 & 4] the CH undergoes a round of refinement, which is crucial to the asymptotic running time of the algorithm. We did not implement these refinements, believing their real-world benefits to be negligible. However, our experiments on *hierarchically structured graphs* (which, in effect, have pre-refined CHs) are very encouraging. They suggest that the refinement step could speed up the computation of shortest paths, at the cost of more pre-computation.

4.3   Breadth First Search

We compare the PR algorithm not only with Dijkstra's, but also with breadth first search (BFS), an effective lower bound on the SSSP problem. Our BFS routine is implemented in the usual way, with a FIFO queue [CLR90]. It finds a shortest path (in terms of number of edges) from the source to all other vertices, and computes the lengths of such paths.

5   Experimental Set-Up

Our main experimental platform was a SunBlade with a 400 MHz clock and 2GB DRAM and a small cache (.5 MB). The large main memory allowed us to test graphs with millions of vertices. For comparison purposes we also ran our code on selected inputs on the following machines.

1. PC running Debian Linux with a 731 MHz Pentium III processor and 255 MB DRAM.
2. SUN Ultra 60 with a 400 MHz clock, 256 MB DRAM, and a 4 MB cache.
3. HP/UX J282 with 180 MHz clock, 128 MB ECC memory.

5.1   Graph Classes

We ran both algorithms on the following classes of graphs.

$G_{n,m}$

The distribution $G_{n,m}$ assigns equal probability to all graphs with m edges on n labeled vertices (see [ER61,Bo85] for structural properties of $G_{n,m}$). We assign edge-lengths identically and independently, using either the uniform distribution over $[0, 1)$, or the *log-uniform* distribution, where edge lengths are given the value 2^q, q being uniformly distributed over $[0, C)$ for some constant C. We use $C = 100$.

Geometric graphs

Here we generate n random points (the vertices) in the unit square and connect with edges those pairs within some specified distance. Edge-lengths correspond to the distance between points. We present results for distance $1.5/\sqrt{n}$, implying an average degree $\approx 9\pi/4$ which is about 7.

Very sparse graphs

These graphs are generated in two stages: we first generate a random spanning tree, to ensure connectedness, then generate an additional $n/10$ random edges. All edges-lengths are uniformly distributed.

Grid graphs

In many situations the graph topology is not random at all but highly predictable. We examine two classes of grid graphs: $\sqrt{n} \times \sqrt{n}$ square grids and $16 \times n/16$ long grids, both with uniformly distributed edge-lengths [CGR96].

New graph classes. Random graphs can have properties that might actually be improbable in real-world situations. For example, $G_{n,m}$ almost surely produces graphs with low diameter, nice expansion properties, and very few small, dense subgraphs [Bo85]. On the other hand, it may be that graph structure is less crucial to the performance of shortest path algorithms than edge length distribution. In the [PR02] algorithm for instance, all the random graph classes described above look almost identical when viewed through the prism of the component hierarchy. They generally produce short hierarchies, where nodes on the lower levels have just a few children and upper level nodes have vast numbers of children.

We introduce two classes of structured random graphs, *Hierarchical* and *Bullseye*, whose component hierarchies are almost predetermined. Edge lengths will be assigned randomly, though according to different distributions depending on how the edge fits into the overall structure.

Hierarchical graphs

These graphs are organized into a hierarchy of clusters, where the lowest level clusters are composed of vertices, and level i clusters are just composed of level $i - 1$ clusters. A hierarchical graph is parameterized by the branching factor b and is constructed so that the CH is almost surely a full b-ary tree of height $\log_b n$. The graph density is also $O(\log_b n)$. We present results for $b = 6$ and $b = 10$.

Bullseye graphs

Bullseye graphs are parameterized by two numbers, the average degree d and the number of *orbits* o. Such a graph is generated by dividing the vertices into o groups of n/o vertices each (the orbits), and assigning $dn/2o$ random edges per orbit, plus a small number of inter-orbit edges to connect the orbits. Edge lengths are assigned depending on the orbits of the endpoints. An intra-orbit edge in orbit i, or an inter-orbit edge where i is the larger orbit is assigned a length uniformly from $[2^i, 2^{i+1})$. The resulting component hierarchy is almost surely a chain of o internal nodes, each with n/o leaf children. We present results for $o = 25$ and $o = 100$ with average degree $d = 3$.

6   Results

The plots in Figures (a)-(j) give the running times of the two algorithms and BFS on the SunBlade for each of the graph classes we considered. Each point in the plots represents the time to compute SSSP/BFS, averaged over thirty trials from randomly chosen sources, on three randomly chosen graphs from the class. The y-axis is a measure of 'microseconds per edge', that is, the time to perform one SSSP/BFS computation divided by the number of edges.

In the plots, DIJ stands for the cost of computing SSSP using Dijkstra's algorithm and PR-marg stands for the *marginal cost* of computing SSSP using the Pettie-Ramachandran algorithm. By marginal cost for the PR algorithm, we mean *the time to compute SSSP after the CH is constructed, and excluding the cost of computing the CH.*

It is unclear how close a *robust* SSSP algorithm can get to the speed of BFS. Our results show that on a variety of graph types the marginal cost of the PR

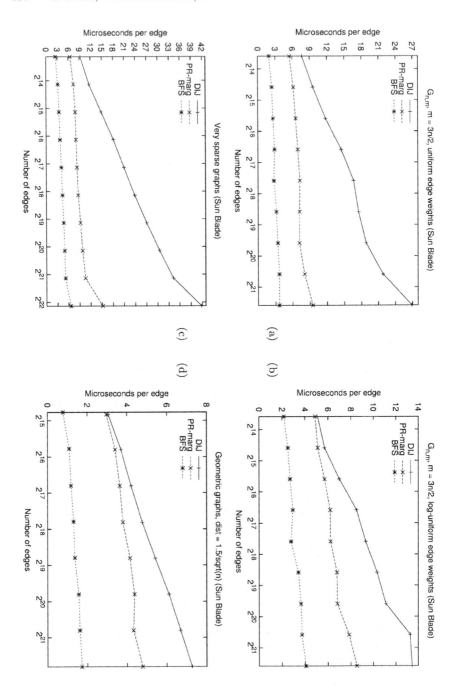

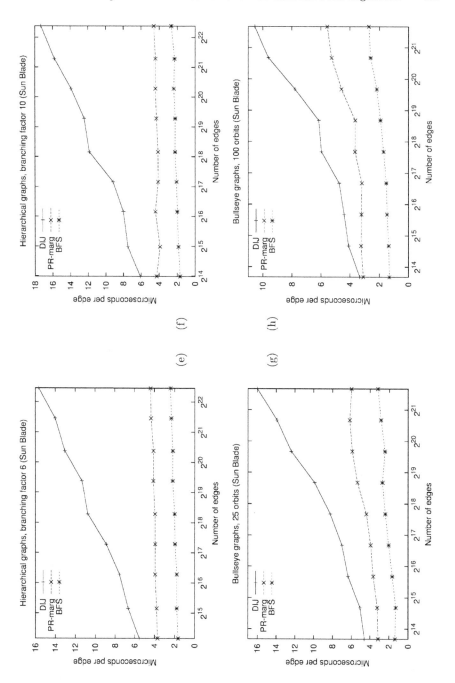

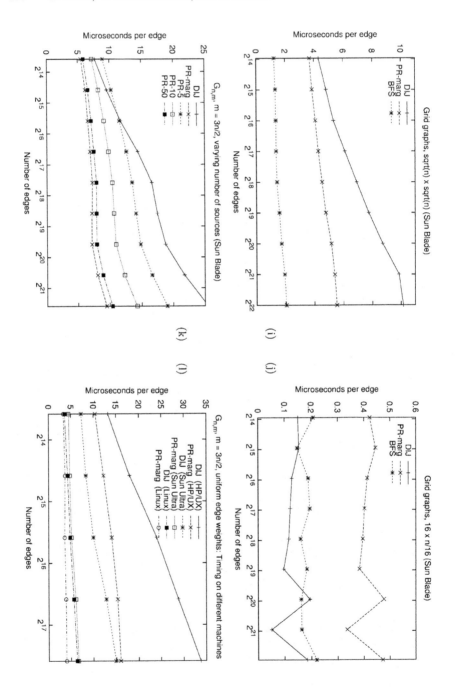

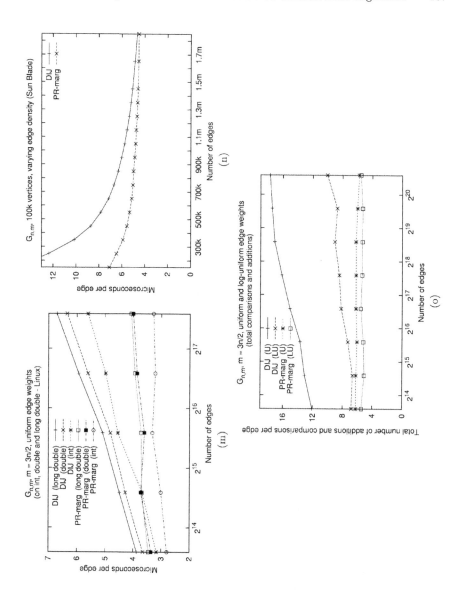

algorithm is *very* competitive with BFS, running between a factor of 1.87 and 2.77 times the BFS speed and usually less than 2.15 times BFS speed — see Table 1, third row. Naturally there is always room for some improvement; the question is, how much slack is there in the PR algorithm?

The effect of the CH pre-computation time is described in Table 1 and Figure (k). Table 1 (first row) lists, for each class of graphs, the critical number of sources s_0 such that PR (including the cost of computing CH) outperforms Dijkstra when SSSP is solved from at least s_0 sources on a graph with 2^{20} vertices. Figure (k) shows the amortized time per SSSP, *including* the cost of computing the CH, for varying numbers of sources in the $G_{n,m}$ graph class. Table 1 indicates that PR overtakes DIJ for a modest number of sources (on all but the long grid graph class), and Figure (k) indicates that the pre-computation time quickly becomes negligible as the number of sources increases. In Figure (k), the line PR-i represents the amortized cost per source (including pre-computation) when computing SSSP from i sources.

Table 1. *First line: number of SSSP computations (s_0) beyond which PR (including cost of computing CH) outperforms Dijkstra. Second line: ratio of time to construct the CH to the time for PR to perform one marginal SSSP computation. Third line: ratio of time for PR to perform one marginal SSSP computation to time for one BFS computation. These statistics reflect graphs of 2^{20} vertices.*

	$G_{n,m}$ (uniform)	$G_{n,m}$ (log-uni.)	Sparse	Geom.	Hier. $b=6$	Hier. $b=10$	Bull. $o=25$	Bull. $o=100$	Grid (sq.)	Grid (long)
s_0	3	21	3	17	4	4	4	7	10	n/a
CH/PR	5.05	11.75	4.38	8.48	10.1	9.6	5.33	6.11	7.85	82.66
PR/BFS	2.14	2.11	1.99	2.77	1.87	1.92	2.15	2.01	2.77	2.06

Figures (a) and (b) show the marginal performance of the PR algorithm to be stable over the uniform and log-uniform distributions. What is somewhat surprising is that Dijkstra's algorithm is dramatically faster under the log-uniform distribution (though still somewhat slower than the marginal performance of the PR algorithm for the same class). We hypothesize that this effect is due to the pairing heap. Recently, Iacono [Iac00] proved that the amortized complexity of extract-min in a pairing heap is logarithmic in the number of operations since the extracted item's insertion. Assigning edge lengths from the log-uniform distribution seems to cause Dijkstra's algorithm to visit vertices which were recently inserted into the heap. An interesting experiment would be to use a heap less sensitive to edge-length distribution, such as a binary heap. The plot for very sparse graphs in Figure (c) shows a nice separation between the marginal cost of the PR algorithm and the cost of Dijkstra's algorithm.

Figure (d), on geometric graphs, still shows the marginal cost of PR to be faster than Dijkstra on all graphs tested, though the separation in running times is not as dramatic as in $G_{n,m}$. We believe this is largely due to the density of the graph (the average degree for the graphs tested in Figure (e) is over 7) and the overhead for relaxing edges in PR, which is worse than in Dijkstra's algorithm. Another factor which could be taken into account is the large diameter of the geometric graphs, which is almost always $\Omega(\sqrt{n})$.

We believe the near-linear marginal costs of PR on Hierarchical graphs (Figures (e) and (f)) are a good indication of how well the *full* PR algorithm [PR02] could perform on all the graph classes. This is due to the structure of the component hierarchy. The CH which is derived naturally from a Hierarchical graph is very similar to the CH of *any* graph which is derived using the refinement step in [PR02, Sections 3 & 4]. The results for the Bullseye graphs are similar to those for $G_{n,m}$ for uniform and log-uniform distributions — DIJ performs better when the number of orbits increases.

The only graph class for which Dijkstra's algorithm beat the marginal cost of PR was the 'long grid' [CGR96], shown in Figure (j). This is to be expected. At any given moment in Dijkstra's algorithm the heap probably contains a constant number of elements, hence the worst-case $n \log n$ term in Dijkstra's algorithm never appears. The running times of the algorithms on long grids appear jumpy because of some aberrant delays which affect a small portion of the SSSP/BFS computations. In the case of square grids, Figure (i), Dijkstra's algorithm did exhibit a super-linear running time. The grid topology of the graph did not seem to have any unpredictable effect on either algorithm.

The results on the SUN Ultra 60, the Linux PC and the HP/UX machines are similar (see figure (l) for a comparison of runs on $G_{n,m}$ with uniform distribution of edge lengths), except that the runs are much faster on the Linux PC and much slower on the HP machine. The Linux PC was also much more sensitive to whether edge-lengths are integers, or floating points with double precision, or floating points with quadruple precision (see Figure (m)). In contrast the results on the SUN machines were virtually the same for integers and double-precision floating points. Note that we needed to alter just one line of code to move between ints, doubles, and long doubles. This is one advantage of the comparison-addition model.

Figure (n) shows the change in running time of both algorithms as the number of edges is increased for a fixed number of vertices. This study was performed on $G_{n,m}$ with 100,000 vertices and with uniform distribution of edge lengths. Dijkstra's algorithm seems to converge to a linear running time as the edge density increases. However, the figure shows the marginal cost of the PR algorithm to be slightly superior even for relatively large edge densities.

Finally in Figure (o) we plot the comparison-addition cost of our implementation of Dijkstra's algorithm and the comparison-addition marginal cost of our implementation of the PR algorithm. The plots are for $G_{n,m}$ with uniform and log-uniform edge length distribution. It is interesting to note that this cost appears to be practically linear for both types of graphs for PR while it is super-linear for DIJ. This plot once again shows up the significantly better performance of DIJ on log-uniform distribution over uniform distribution of edge lengths. These results are, of course, specific to the design choices we made for our implementations (in particular, the use of the pairing heap).

7   Discussion

We have implemented a simplified version of the Pettie-Ramachandran shortest path algorithm for undirected graphs and tested it against its chief competitor: Dijkstra's algorithm. The evidence shows that the pre-computation time of the PR algorithm is time well spent if we proceed to compute multiple-source shortest paths from enough sources.

We did not compare our algorithm directly to any integer-based shortest path algorithms, the focus of [CGR96,GS97,G01b], however we do compare it against breadth first search, a practical lower bound on the shortest path problem. In Goldberg's [G01b] recent study, the best algorithms performed (roughly) between 1.6 and 2.5 times the BFS speed,[5] whereas the PR algorithm performed 1.87 to 2.77 times slower than BFS, a remarkable fact considering the algorithms tested in [G01b] were specifically engineered for small integer edge lengths.

One issue we did not directly address is whether the PR algorithm's gain in speed is due to caching effects, or whether it is genuinely performing fewer operations than Dijkstra's algorithm. The data on comparison/addition operations[6] versus running time data suggests that the cache miss-rate is roughly equal in both Dijkstra's algorithm and the PR algorithm. We suspect that plugging in a cache-sensitive heap, such as [S00], will affect the performance of both algorithms similarly.

An open problem is to develop a shortest path algorithm for undirected graphs which beats Dijkstra's when computing *single*-source shortest paths on sparse graphs. We think the component hierarchy approach can lead to such an algorithm (and a qualified success appears in [PR02]). However, the possibility of a *practical* SSSP algorithm based on the component hierarchy is unlikely since it requires computing the MST in advance, and the experimental results in [MS94] suggest that the fastest method (in practice) for computing MST is the Prim-Dijkstra algorithm — which is structured nearly identically to Dijkstra's SSSP algorithm [Dij59].

It would be interesting to see if the performance of PR could be improved by using the hierarchical bucketing structure (using only comparisons and additions) assumed in [PR02] rather than the pairing heap used in our experiments. Very similar bucketing structures were used in two recent theoretical SSSP algorithms [Mey01,G01], both with good *average-case* performance. Both assume uniformly distributed edge lengths. An open question is whether either of these algorithms work well in practice (and if they are competitive with [PR02]), and how sensitive each is to edge-length distribution.

---

[5] Goldberg implements his BFS with *the same data structures* used in one of the algorithms, which, if slower than the usual BFS, would bias the timings.

[6] The number of comparisons/additions in PR and DIJ closely correspond to the total number of operations.

References

[AHU74] A. V. Aho, J. E. Hopcroft, J. D. Ullman. *The Design and Analysis of Computer Algorithms*. Addison-Wesley, 1974.

[Bo85] B. Bollobás. *Random Graphs*. Academic Press, London, 1985.

[CGR96] B. V. Cherkassky, A. V. Goldberg, T. Radzik. Shortest paths algorithms: Theory and experimental evaluation. In *Math. Prog.* 73 (1996), 129-174.

[CLR90] T. Cormen, C. Leiserson, R. Rivest. *Intro. to Algorithms*. MIT Press, 1990.

[Dij59] E. W. Dijkstra. A note on two problems in connexion with graphs. In *Numer. Math.*, 1 (1959), 269-271.

[ER61] P. Erdös, A. Rényi On the evolution of random graphs. *Bull. Inst. Internat. Statist.* 38, pp. 343–347, 1961.

[F76] M. Fredman. New bounds on the complexity of the shortest path problem. *SIAM J. Comput.* 5 (1976), no. 1, 83–89.

[F+86] M. L. Fredman, R. Sedgewick, D. D. Sleator, R. E. Tarjan. The pairing heap: A new form of self-adjusting heap. In *Algorithmica* 1 (1986) pp. 111-129.

[FT87] M. L. Fredman, R. E. Tarjan. Fibonacci heaps and their uses in improved network optimization algorithms. In *JACM* 34 (1987), 596–615.

[G85] H. N. Gabow. A scaling algorithm for weighted matching on general graphs. In *Proc. FOCS 1985*, 90–99.

[G01] A. Goldberg. A simple shortest path algorithm with linear average time. InterTrust Technical Report STAR-TR-01-03, March 2001.

[G01b] A. Goldberg. Shortest path algorithms: engineering aspects. *ISSAC 2001*.

[GS97] A. Goldberg, C. Silverstein. Implementations of Dijkstra's algorithm based on multi-level buckets. *Network optimization* (1997), Lec. Not. Econ. Math. Syst. 450, 292–327.

[Hag00] T. Hagerup. Improved shortest paths on the word RAM. In *Proc. ICALP 2000*, LNCS volume 1853, 61–72.

[Iac00] J. Iacono. Improved upper bounds for pairing heaps. Algorithm theory— SWAT 2000 (Bergen), LNCS vol. 1851, 32–45,

[Jak91] H. Jakobsson, Mixed-approach algorithms for transitive closure. In *Proc. ACM PODS*, 1991, pp. 199-205.

[Jar30] V. Jarník. O jistém problému minimálním. *Práca Moravské Přírodovědecké Společnosti* 6 (1930), 57–63, in Czech.

[KKP93] D. R. Karger, D. Koller, S. J. Phillips. Finding the hidden path: time bounds for all-pairs shortest paths. *SIAM J. on Comput.* 22 (1993), no. 6, 1199–1217.

[KS98] S. Kolliopoulos, C. Stein. Finding real-valued single-source shortest paths in $o(n^3)$ expected time. *J. Algorithms* 28 (1998), no. 1, 125–141.

[McG91] C. C. McGeoch. A new all-pairs shortest-path algorithm. Tech. Report 91-30 DIMACS, 1991. Also appears in *Algorithmica*, 13(5): 426-461, 1995.

[MN99] K. Mehlhorn, S. Näher. *The LEDA Platform of Combinatorial and Geometric Computing*. Cambridge Univ. Press, 1999.

[Mey01] U. Meyer. Single source shortest paths on arbitrary directed graphs in linear average-case time. In *Proc. SODA 2001*, 797–806.

[MT87] A. Moffat, T. Takaoka. An all pairs shortest path algorithm with expected time $O(n^2 \log n)$. *SIAM J. Comput.* 16 (1987), no. 6, 1023–1031.

[MS94] B. M. E. Moret, H. D. Shapiro. An empirical assessment of algorithms for constructing a minimum spanning tree. In *DIMACS Series on Discrete Math. and Theor. CS*, 1994.

[PR00]   S. Pettie, V. Ramachandran. An optimal minimum spanning tree algorithm. In *Proc. ICALP 2000*, LNCS volume 1853, 49–60. *JACM*, to appear.

[PR02]   S. Pettie, V. Ramachandran. Computing shortest paths with comparisons and additions. In *Proc. SODA '02*, January 2002, to appear.

[S00]   P. Sanders. Fast priority queues for cached memory. *J. Experimental Algorithms* 5, article 7, 2000.

[Tak92]   T. Takaoka. A new upper bound on the complexity of the all pairs shortest path problem. *Inform. Process. Lett.* 43 (1992), no. 4, 195–199.

[Tho99]   M. Thorup. Undirected single source shortest paths with positive integer weights in linear time. *J. Assoc. Comput. Mach.* 46 (1999), no. 3, 362–394.

[Tho01]   M. Thorup. Quick k-median, k-center, and facility location for sparse graphs. In *Proc. ICALP 2001*, LNCS Vol. 2076, 249–260.

[Z01]   U. Zwick. Exact and approximate distances in graphs – a survey. In *Proc. 9th ESA* (2001), 33–48. Updated copy at `http://www.cs.tau.ac.il/~zwick`

Getting More from Out-of-Core Columnsort

Geeta Chaudhry[*] and Thomas H. Cormen

Dartmouth College Department of Computer Science
{geetac,thc}@cs.dartmouth.edu

Abstract. We describe two improvements to a previous implementation of out-of-core columnsort, in which data reside on multiple disks. The first improvement replaces asynchronous I/O and communication calls by synchronous calls within a threaded framework. Experimental runs show that this improvement reduces the running time to approximately half of the running time of the previous implementation. The second improvement uses algorithmic and engineering techniques to reduce the number of passes over the data from four to three. Experimental evidence shows that this improvement yields modest performance gains. We expect that the performance gain of this second improvement increases when the relative speed of processing and communication increases with respect to disk I/O speeds. Thus, as processing and communication become faster relative to I/O, this second improvement may yield better results than it currently does.

1 Introduction

In a previous paper [1], the authors reported on an out-of-core sorting program based on Leighton's columnsort algorithm [2]. By some resource measures—specifically, disk time and processor time plus disk time—our columnsort-based algorithm was more sorting-efficient than the renowned NOW-Sort program [3].[1] Unlike NOW-Sort, the implementation of our algorithm performed interprocessor communication and disk I/O using only standard, off-the-shelf software, such as MPI [4] and MPI-2 [5].

The present paper explores two improvements to the implementation of our out-of-core program:

1. Overlapping I/O, computation, and communication by means of a threaded implementation in which all calls to MPI and MPI-2 functions are synchronous. Asynchrony is provided by the standard pthreads package. The previous implementation achieved asynchrony by making calls to the asynchronous versions of MPI and MPI-2 functions.

---

[*] Geeta Chaudhry was supported in part by a donation from Sun Microsystems. This research was supported in part by NSF Grant EIA-98-02068.

[1] For two other measures—specifically, processor time and used-memory time—NOW-Sort was more sorting-efficient than our columnsort-based algorithm. Also, for processor time plus disk time, our algorithm was more sorting-efficient than NOW-Sort in only some cases.

D. Mount and C. Stein (Eds.): ALENEX 2002, LNCS 2409, pp. 143–154, 2002.
© Springer-Verlag Berlin Heidelberg 2002

2. Reducing the number of passes over the data from four down to three. An algorithmic observation makes this reduction possible.

We shall refer to our prior implementation as the *non-threaded 4-pass* implementation, and to our new implementations as the *threaded 4-pass* and *threaded 3-pass* implementations. We shall see that the threaded 4-pass implementation can reduce the overall running time down to approximately half of the non-threaded 4-pass implementation's running time. Moreover, the threaded 4-pass implementation allows greater flexibility in memory usage by both the user and the program itself. The effect of reducing the number of passes is more modest; experiments show that the running time of the threaded 3-pass implementation is between 91.5% and 94.6% of that of the threaded 4-pass implementation.

The remainder of this paper is organized as follows. Section 2 summarizes columnsort and presents the out-of-core algorithm originally described in [1]. Section 3 outlines the differences between the non-threaded 4-pass implementation with asynchronous MPI and MPI-2 calls and the threaded 4-pass implementation. In Section 4, we describe our threaded 3-pass implementation. We give empirical results for all implementations. Finally, Section 5 offers some concluding remarks.

2   Background

In this section, we review our non-threaded 4-pass implementation of columnsort from the previous paper [1]. After presenting the basic columnsort algorithm, we describe its adaptation to an out-of-core setting. We conclude this section with a discussion of the performance results.

The Basic Columnsort Algorithm. Columnsort sorts N *records*. Each record contains a *key*. In columnsort, the records are arranged into an $r \times s$ matrix, where $N = rs$, s is a divisor of r, and $r \geq 2(s-1)^2$. When columnsort completes, the matrix is sorted in column-major order. That is, each column is sorted, and the keys in each column are no larger than the keys in columns to the right.

Columnsort proceeds in eight steps. Steps 1, 3, 5, and 7 are all the same: sort each column individually. Each of steps 2, 4, 6, and 8 permutes the matrix entries as follows:

Step 2: Transpose and reshape: We first transpose the $r \times s$ matrix into an $s \times r$ matrix. Then we "reshape" it back into an $r \times s$ matrix by taking each row of r entries and rewriting it as an $r/s \times s$ submatrix. For example, the column with $r = 6$ entries $a\ b\ c\ d\ e\ f$ is transposed into a 6-entry row with entries $a\ b\ c\ d\ e\ f$ and then reshaped into the 2×3 submatrix $\begin{bmatrix} a & b & c \\ d & e & f \end{bmatrix}$.

Step 4: Reshape and transpose: We first reshape each set of r/s rows into a single r-element row and then transpose the matrix. This permutation is the inverse of that of step 2.

Step 6: Shift down by $r/2$: We shift each column down by $r/2$ positions, wrapping into the next column. That is, we shift the top half of each column into the bottom half of that column, and we shift the bottom half of each column into the top half of the next column.

Step 8: Shift up by $r/2$: We shift each column up by $r/2$ positions, wrapping around into the previous column. This permutation is the inverse of that of step 6.

Our Out-of-Core Columnsort. Our adaptation of columnsort to an out-of-core setting assumes that the machine has P processors $\mathcal{P}_0, \mathcal{P}_1, \ldots, \mathcal{P}_{P-1}$ and D disks $\mathcal{D}_0, \mathcal{D}_1, \ldots, \mathcal{D}_{D-1}$. When $D = P$, each processor accesses exactly one disk over the entire course of the algorithm. When $D < P$, we require that there be P/D processors per node and that they share the node's disk; in this case, each processor accesses a distinct portion of the disk. In fact, in our implementation, we treat this distinct portion as a separate "virtual disk," allowing us to assume that $D \geq P$. When $D > P$, each processor has exclusive access to D/P disks. We say that a processor *owns* the D/P disks that it accesses.

We use buffers that hold exactly r records. Each processor has several such buffers. For convenience, our current implementation assumes that all parameters (including r) are powers of 2.

There is an upper limit on the number of records N in the file to be sorted. Recalling that $N = rs$ and that each buffer of r records must fit in the memory of a single processor, this limit occurs because of the columnsort requirement that $r \geq 2(s - 1)^2$. If we simplify this requirement to $r \geq 2s^2$, we have the restriction that $r \geq 2(N/r)^2$, which is equivalent to $N \leq r^{3/2}/\sqrt{2}$.

The data are placed so that each column is stored in contiguous locations on the disks owned by a single processor. Specifically, processor j *owns* columns $j, j + P, j + 2P$, and so on.

Here, we outline the basic structure of each pass of our non-threaded 4-pass implementation along with the key implementation features; for details, see [1]. Each pass performs two consecutive steps of columnsort. That is, pass 1 performs columnsort steps 1 and 2, pass 2 performs steps 3 and 4, pass 3 performs steps 5 and 6, and pass 4 performs steps 7 and 8. Each pass is decomposed into s/P *rounds*. Each round processes the next set of P consecutive columns, one column per processor, in five phases:

Read phase: Each processor reads a column of r records from the disks that it owns. We use asynchronous MPI-2 calls to implement this phase.

Sort phase: Each processor locally sorts, in memory, the r records it has just read. Implementation of this phase differs from pass to pass. Pass 1 uses the system `qsort` call. In the other passes, the data starts with some sorted runs, depending on the write pattern of the previous pass. In passes 2 and 3, each processor has s sorted runs of size r/s each, and so we use recursive merging with $\lg s$ levels. In pass 4, each processor starts with two sorted runs of size $r/2$ each, and so local sorting requires just one level of merging to obtain a single sorted run.

Communicate phase: Each record is destined for a specific column, depending on which even-numbered columnsort step this pass is performing. In order to get each record to the processor that owns this destination column, processors exchange records. We use asynchronous MPI calls to perform this communication. In passes 2, 3, and 4, this phase requires every processor to send some data to every other processor. In pass 3, the communication is simpler, since every processor communicates with only two other processors.

Permute phase: Having received records from other processors, each processor rearranges them into the correct order for writing.

Write phase: Each processor writes a set of records onto the disks that it owns. These records are not necessarily all written consecutively onto the disks, though they are written as a small number of sorted runs. Again, we use asynchronous MPI-2 calls to perform the writes.

Note that the use of asynchronous MPI and MPI-2 calls allows us to overlap local sorting, communication, and I/O. At any particular time, processor j might be communicating records belonging to column $j+kP$, locally sorting records in column $j+(k+1)P$, reading column $j+(k+2)P$, and writing column $j+(k-1)P$. However, our non-threaded 4-pass implementation does not overlap reading with either local sorting or writing.

Performance Results. Table 1 summarizes the performance results of several implementations of out-of-core columnsort.[2] Our performance goal is to sort large volumes of data while consuming as few resources as possible.

The results are for 64 byte records with an integer key at the beginning of each record. The input files are generated using the drand function to generate the value of each key.

The results in Table 1 are for two different clusters of SMPs. The first system is a cluster of 4 Sun Enterprise™ 450 SMPs. We used only one processor on each node. Each processor is an UltraSPARC™-II, running at 296 MHz and with 128 MB of RAM, of which we used 40 MB for holding data. The nodes are connected by an ATM OC-3 network, with a peak speed of 155 megabits per second. This system has one disk per node, each an IBM DNES309170 spinning at 7200 RPM and with an average latency of 4.17 msec. All disks are on Ultra2 SCSI buses. The MPI and MPI-2 implementations are part of Sun HPC ClusterTools 4™.

The second system is also a cluster of 4 Sun Enterprise 450 SMPs. Here, we used two processors on each node. Other than the disks, the components are the same as in the first system. This system has 8 disks, 4 of which are the same IBM disks as in the first system, and the other 4 of which are Seagate ST32171W also spinning at 7200 RPM and also with an average latency of 4.17 msec.

We conclude this section by discussing two additional features of our implementations:

- Our algorithm makes no assumptions about the keys. In fact, our algorithm's I/O and communication patterns are oblivious to the keys. That is, the exact

---

[2] The results in the columns labeled "non-threaded 4-pass" are for more recent runs than in [1].

Table 1. Parameters and results for three implementations of our out-of-core column-sort algorithm on two clusters of Sun Enterprise 450 SMPs. The rows show number of processors (P), number of disks (D), megabytes of memory used across the entire system (U), gigabytes sorted (V), and time to sort (T) for each system. Times shown are averages of five runs.

	non-threaded 4-pass	threaded 4-pass	threaded 3-pass	non-threaded 4-pass	threaded 4-pass	threaded 3-pass
P	4	4	4	8	8	8
D	4	4	4	8	8	8
U (MB)	160	192	256	320	384	512
V (GB)	1	1	1	2	2	2
T (minutes)	4.338	2.161	2.043	5.631	3.217	2.944

sequence of I/O and communication operations depend only on the problem size and the system parameters, not on the key values. Because the I/O and communication patterns are completely oblivious to the keys, each round has no effect at all on other rounds of the same pass. That is, the rounds of a pass are fully independent of each other and therefore can execute with full asynchrony. We shall take advantage of this property in our threaded implementations.

– Our algorithm produces sorted records in the standard striped ordering used by the Parallel Disk Model (PDM). This ordering affects only the write phase of the last pass, in which each column is striped across all D disks rather than residing solely on the disks of the processor that owns it. PDM ordering balances the load for any consecutive set of records across processors and disks as evenly as possible. A further advantage to producing sorted output in PDM ordering is that our algorithm can be used as a subroutine in other PDM algorithms. To the best of our knowledge, our prior implementation is the first multiprocessor sorting algorithm whose output is in PDM order.

3   Threaded 4-Pass Implementation

In this section, we describe our threaded 4-pass implementation and compare its performance to that of the non-threaded 4-pass implementation.

Motivation. There are two reasons to revise our implementation to use threads. First, threads allow scheduling to be more dynamic. Second, our threaded implementation permits greater flexibility in memory usage.

Our columnsort algorithm uses four types of machine resources: disks, the communication network, processing power, and memory. As described in the previous section, each column goes through five phases, two using I/O (the read and write phases), one using the network (the communicate phase), and the other two using the processing power (the sort and permute phases). All phases make use of memory, of course.

Our non-threaded 4-pass implementation has a single thread of control. In order to permit overlapping of I/O, communication, and processing, it relies on asynchronous MPI and MPI-2 calls in order to schedule the usage of the four types of resources. Due to the single thread of control, this implementation has to decide the overlap mechanism statically at coding time and is not fully flexible. For example, the implementation is unable to adapt automatically to a faster network, faster CPU, or faster I/O. Our threaded 4-pass implementation uses the standard pthreads package to overlap I/O, communication, and processing. It uses only synchronous MPI and MPI-2 calls.

Because the non-threaded 4-pass implementation uses static scheduling, its memory usage is static and has to be known at coding time. Consequently, this implementation is unable to adapt to an increased amount of available memory. Our threaded 4-pass implementation, on the other hand, maintains a global pool of memory buffers, the number of which is set at the start of each run of the program. One can determine the optimum number of buffers for a given configuration by means of a small number of experimental runs.

Basic Structure. In order to overlap the usage of resources in a dynamic manner, we created four threads per processor. One thread is responsible for all the disk I/O functions, one does all the interprocessor communication, one does all the sorting, and the final thread does all the permuting. We shall refer to the four threads as the I/O, communicate, sort, and permute threads, respectively. The threads operate on buffers, each capable of holding exactly r records, and which are drawn from a global pool. The threads communicate with each other via a standard semaphore mechanism. The read and write functions appear in a common thread—the I/O thread—because they will serialize at the disk anyway.

Figure 1 illustrates how the threads work by displaying the history of a column within a given round of a given pass.

1. The I/O thread acquires a buffer b from the global pool and performs the read phase of column c by reading the column from the disk into this buffer. The I/O thread suspends while the read happens, and when the read completes, the I/O thread wakes up.

2. When the I/O thread wakes up after the read completes, it signals the sort thread, which picks up buffer b and performs the sort phase of column c on it.

3. The sort thread signals the communicate thread, which picks up buffer b and performs the communicate phase of column c on it. The communicate phase suspends during interprocessor communication, and after communication completes, it wakes up.

4. The communicate thread signals the permute thread, which picks up buffer b and performs the permute phase of column c on it.

5. Finally, after the permute phase completes, the permute thread signals the I/O thread, which picks up buffer b and writes it out to disk. The I/O thread suspends during the write, and when the write completes, the I/O thread wakes up and releases buffer b back to the global pool.

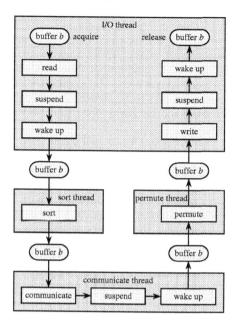

Fig. 1. The history of a buffer b as it progresses within a given round of a given pass. The I/O thread acquires the buffer from the global pool and then reads into it from disk. The I/O thread suspends during the read, and when it wakes up, it signals the sort thread. The sort thread sorts buffer b and signals the communicate thread. The communicate thread suspends during interprocessor communication, and when it wakes up, it signals the permute thread. The permute thread then permutes buffer b and signals the I/O thread. The I/O thread writes the buffer to disk, suspending during the write. When the I/O thread wakes up, it releases buffer b back to the global pool.

The pthreads implementation may preempt any thread at any time. Thus, during the time that a given thread considers itself as active, it might not actually be running on the CPU.

The sort, permute, and communicate threads allocate additional buffers for their own use. Each of these threads allocates one buffer at the beginning of the program and uses it throughout the entire run. Thus, the total memory usage of the threaded 4-pass implementation is three buffers more than are created in the global pool.

Performance Results. From the columns labeled "threaded 4-pass" in Table 1, we see that the threaded 4-pass implementation takes only 49.8% of the time taken by the non-threaded 4-pass implementation on the cluster with 4 processors and 4 disks. On the cluster with 8 processors and 8 disks, the threaded 4-pass implementation takes 57.1% as much time as the non-threaded 4-pass implementation.

What accounts for this significant improvement in running time? Due to the highly asynchronous nature of each of the implementations, we were unable to obtain accurate breakdowns of where any of them were truly spending their time. We were able to obtain the amounts of time that each thread considered itself as active, but because threads may be preempted, these times may not be reflective of the times that the threads were actually running on the CPU. Similarly, the timing breakdown for the non-threaded 4-pass implementation is not particularly accurate.

Our best guess is that the gains come primarily from two sources. First is increased flexibility in scheduling, which we discussed above in regard to the motivation for threaded implementations. The second source of performance gain is that the MPI calls in the threaded 4-pass implementation are synchronous, whereas they are asynchronous in the non-threaded 4-pass implementation. Apparently, asynchronous MPI calls incur a significant overhead. Although there is an overhead cost due to threads, the benefits of scheduling flexibility and synchronous MPI calls in the threaded 4-pass implementation outweigh this cost.

We conducted a set of ancillary tests to verify that a program with threads and synchronous MPI calls is faster than a single-threaded program with asynchronous MPI calls. The first test overlaps computation and I/O, and the second test overlaps computation and communication. We found that by converting a single thread with asynchronous MPI calls to a threaded program with synchronous MPI calls, the computation-and-I/O program ran 7.8% faster and the computation-and-communication program ran 23.8% faster.

Moreover, there is a qualitative benefit of the threaded 4-pass implementation. Because all calls to MPI and MPI-2 functions are synchronous, the code itself is cleaner, and it is easier to modify.

4   Threaded 3-Pass Implementation

This section describes how to reduce the number of passes in the implementation given in the previous section from four to three. The key observation is the *pairing observation* from [1]:

> We can combine steps 6–8 of columnsort by pairing adjacent columns. We sort the bottom $r/2$ entries of each column along with the top $r/2$ entries of the next column, placing the sorted r entries into the same positions. The top $r/2$ entries of the leftmost column were already sorted by step 5 and can therefore be left alone, and similarly for the bottom $r/2$ entries of the rightmost column.

Basic Structure. To take advantage of the pairing implementation, we combine steps 5–8 of columnsort—passes 3 and 4 in a 4-pass implementation—into one pass. Figure 2 shows how. In the 4-pass implementation, the communicate, permute, and write phases of pass 3, along with the read phase of pass 4, merely shift each column down by $r/2$ rows (wrapping the bottom half of each column into the top half of the next column). We replace these four phases by a single communicate phase.

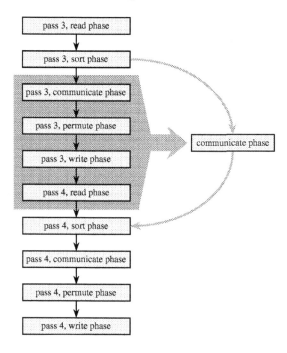

Fig. 2. How passes 3 and 4 of the threaded 4-pass implementation are combined into a single pass of the threaded 3-pass implementation. The communicate, permute, and write phases of pass 3, and the read phase of pass 4 are replaced by a single communicate phase.

In the threaded 3-pass implementation, the first two passes are the same as in the threaded 4-pass implementation. To further understand how the last pass works, let us examine a typical round (i.e., neither the first nor the last round) in this pass. At the start of the round, some processor $\mathcal{P}_k$ contains in a buffer $r/2$ records left over from the previous round. When the round completes, some other processor $\mathcal{P}_l$, where $l = (k-1) \bmod P$, will contain $r/2$ leftover records, and it will serve as $\mathcal{P}_k$ in the next round. The round proceeds as follows:

1. Each processor except for $\mathcal{P}_k$ reads in a column, so that $P - 1$ columns are read in.
2. Each processor except for $\mathcal{P}_k$ sorts its column locally.
3. With the exception of $\mathcal{P}_k$, each processor $\mathcal{P}_i$ sends the first $r/2$ of its sorted elements to processor $\mathcal{P}_{(i-1) \bmod P}$. After all sends are complete, each processor except for $\mathcal{P}_l$ holds r records ($r/2$ that it had prior to the send, and $r/2$ that it just received), and processor $\mathcal{P}_l$ holds $r/2$ records, which it keeps aside in a separate buffer to be used in the next round. This communicate phase replaces the four phases shaded in Figure 2.
4. Each processor except for $\mathcal{P}_l$ sorts its column locally.

5. To prepare for the write in PDM order, each processor (except $\mathcal{P}_l$) sends r/P records to every processor (including $\mathcal{P}_l$ and itself).
6. Each processor locally permutes the $r(P-1)/P$ records it has received to put them into the correct PDM order.
7. Each processor writes the $r(P-1)/P$ records to the disks that it owns.

The first and last rounds have minor differences from the middle rounds. The first round processes all P columns, and the last round may process fewer than $P-1$ columns. Because the first round processes P columns, the number of rounds is $1 + \lceil (s-P)/(P-1) \rceil$.

Thread Structure. In the threaded 3-pass implementation, the first two passes have the same thread structure as in the threaded 4-pass implementation, but the third pass is different. As we have just seen, the third pass has seven phases. As before, each phase is assigned to a single thread, except that the read and write phases are assigned to a single I/O thread. Consequently, there is one I/O thread, two communicate threads, two sort threads, and one permute thread. This thread structure raises two additional issues.

First, the number of additional buffers increases. Other than the I/O thread, each thread requires an r-record buffer. Since there are five non-I/O threads, this implementation requires five buffers more than are in the global pool. That is, the threaded 3-pass implementation requires two buffers more than the threaded 4-pass implementation. The figures for memory used in Table 1 reflect the larger number of buffers required for the threaded 3-pass implementation.

Second, because there are two communicate threads, each making calls to MPI, not all MPI implementations are suitable. Some MPI implementations are unreliable when multiple threads perform communication. Fortunately, the MPI implementations that we have used—on both the Sun cluster and Silicon Graphics Origin 2000 systems—support multiple threads that communicate via MPI calls.

Performance Results. By inspection of Table 1, we see that on the 4-processor, 4-disk and on the 8-processor, 8-disk clusters, the threaded 3-pass implementation takes 94.6% and 91.5%, respectively, of the time used by the threaded 4-pass implementation. The improvement due to reducing the number of passes from four to three is not as marked as that from introducing a threaded implementation.

The observed running times lead to two questions. First, what accounts for the improvement that we see in eliminating a pass? Second, why don't we see more of an improvement?

Compared to the threaded 4-pass implementation, the threaded 3-pass implementation enjoys one significant benefit and one additional cost. Not surprisingly, the benefit is due to less disk I/O. The pass that is eliminated reads and writes each record once, and so the 3-pass implementation performs only 75% as much disk I/O as the 4-pass implementation. This reduced amount of disk I/O is the only factor we know of that accounts for the observed improvement.

The added cost is due to more sort phases and more communicate phases. In the 3-pass implementation, each round contains two sort and two communicate phases, for a total of $2 + 2 \lceil (s - P)/(P - 1) \rceil$ sort phases and the same number of communicate phases. Together, the two rounds of the 4-pass implementation— which are replaced by the last pass of the 3-pass implementation—perform $2s/P$ sort phases and $2s/P$ communicate phases. Since the problem is out-of-core, we have $s > P$, which in turn implies that $2 + 2 \lceil (s - P)/(P - 1) \rceil \geq 2s/P$. Thus, the 3-pass implementation always has more sort and communicate phases than the 4-pass implementation.

The degree to which the 3-pass implementation improves upon the 4-pass implementation depends on the relative speeds of disks, processing, and communication in the underlying system. The 25% reduction in the number of passes does not necessarily translate into a 25% reduction in the overall time. That is because although the combined third pass of the 3-pass implementation writes and reads each record only once, all the other work (communication and sorting) of the two passes still has to be done. Therefore, when the last two passes of the 4-pass implementation are relatively I/O bound, we would expect the 3-pass implementation to be significantly faster than the 4-pass implementation. Conversely, when the last two passes are not I/O bound, the advantage of the 3-pass implementation is reduced. In fact, the 3-pass implementation can even be slower than the 4-pass implementation! On the clusters whose results appear in Table 1, the processing and network speeds are not particularly fast. Our detailed observations of the 4-pass implementation reveal that in the last two passes, the I/O and communication times are close. Hence, these passes are only slightly I/O bound, and therefore we see only a modest gain in the 3-pass implementation.

One would expect that technology will evolve so that the speed of processing and communication networks will increase more rapidly than the speed of disk I/O. If this prediction holds, then the last two passes of the threaded 4-pass implementation will become increasingly I/O bound, and so the relative advantage of the threaded 3-pass implementation will become more prominent.

5   Conclusion

We have seen two ways to improve upon our earlier, non-threaded, 4-pass implementation of out-of-core columnsort. This original implementation had performance results that, by certain measures, made it competitive with NOW-Sort. The two improvements make the implementation even faster.

One can characterize the threaded 4-pass implementation as an engineering effort, which yielded substantially better performance on the two clusters on which it was tested.

On the other hand, the threaded 3-pass implementation has both algorithmic and engineering aspects. On the particular clusters that served as our testbed, the performance gains were modest. We note, however, that if we were to run the threaded 3-pass implementation on a cluster with faster processors and a faster network, we would expect to see a more significant performance improvement. Our future work includes such experimental runs.

We intentionally omitted a non-threaded 3-pass implementation, even though that would have completed all four cases in the space of threaded vs. non-threaded and 4-pass vs. 3-pass implementations. The authors chose to not implement that option because, having seen the significant performance improvement from the threaded implementation, the substantial effort required to produce a non-threaded, 3-pass implementation would not have been worthwhile.

We have identified several other directions for future work. Can we bypass the size limitation given by $N \leq r^{3/2}/\sqrt{2}$? We have two approaches that we plan to investigate. One is to spread the r-record buffers across the memories of all processors, thereby increasing the maximum value of r by a factor of P. This change would allow the maximum problem size to increase with the amount of memory in the entire system, rather than being limited by the amount of memory per processor, as it is now. The second approach is to apply ideas used in the Revsort algorithm [6] to increase the exponent of r from 3/2 to 5/3. Another question is how best to remove the power-of-2 restriction on all parameters. Yet another opportunity for improvement lies in the observation that among all passes, the bulk of the sorting time occurs in pass 1. Might it be possible to not fully sort each column in pass 1 yet maintain the correctness of the overall algorithm?

Acknowledgments. The authors thank Len Wisniewski of Sun Microsystems, who was exceptionally helpful in preparing our experimental results. Thanks also to the anonymous reviewers for several helpful remarks.

References

1. Chaudhry, G., Cormen, T.H., Wisniewski, L.F.: Columnsort lives! An efficient out-of-core sorting program. In: Proceedings of the Thirteenth Annual ACM Symposium on Parallel Algorithms and Architectures. (2001) 169–178
2. Leighton, T.: Tight bounds on the complexity of parallel sorting. IEEE Transactions on Computers **C-34** (1985) 344–354
3. Arpaci-Dusseau, A.C., Arpaci-Dusseau, R.H., Culler, D.E., Hellerstein, J.M., Patterson, D.A.: High-performance sorting on networks of workstations. In: SIGMOD '97. (1997)
4. Snir, M., Otto, S.W., Huss-Lederman, S., Walker, D.W., Dongarra, J.: MPI—The Complete Reference, Volume 1, The MPI Core. The MIT Press (1998)
5. Gropp, W., Huss-Lederman, S., Lumsdaine, A., Lusk, E., Nitzberg, B., Saphir, W., Snir, M.: MPI—The Complete Reference, Volume 2, The MPI Extensions. The MIT Press (1998)
6. Schnorr, C.P., Shamir, A.: An optimal sorting algorithm for mesh connected computers. In: Proceedings of the Eighteenth Annual ACM Symposium on Theory of Computing. (1986) 255–263

Topological Sweep in Degenerate Cases

Eynat Rafalin[1], Diane Souvaine[1], and Ileana Streinu[2]

[1] Department of Electrical Engineering and Computer Science, Tufts University,
Medford, MA 02155.*** {erafalin, dls}@eecs.tufts.edu
[2] Department of Computer Science, Smith College, Northampton, MA 01063.[†]
streinu@cs.smith.edu

Abstract. Topological sweep can contribute to efficient implementations of various algorithms for data analysis. Real data, however, has degeneracies. The modification of the topological sweep algorithm presented here handles degenerate cases such as parallel or multiply concurrent lines without requiring numerical perturbations to achieve general position. Our method maintains the $O(n^2)$ and $O(n)$ time and space complexities of the original algorithm, and is robust and easy to implement. We present experimental results.

1 Introduction

Dealing with degenerate data is a notoriously untreated problem one has to face when implementing Computational Geometry algorithms. Most of the theoretical developments have avoided the special cases or proposed too general solutions that may produce other side-effects. One of the few papers that actively deals with degenerate cases is Burnikel *et al.* [2]. The authors argue forcefully that perturbation is not always effective in practice and that it is simpler (in terms of programming effort) and more efficient (in terms of running time) to deal directly with degenerate inputs. Their paper presents two implementations for solving basic problems in computational geometry. Note that the running time of their implementation can be sensitive to the amount of degeneracy.

The work in this paper is motivated by the practical need to have a robust implementation of topological sweep to be used within implementations of several geometric algorithms for computing statistical measures of data depth, both those already coded (e.g. [9]) and those currently in development. The implementation is general enough to be used in place of any topological sweep subroutine in existing code.

The *topological sweep* method of Edelsbrunner and Guibas [5] is one of the classical algorithms in Computational Geometry. It sweeps an arrangement of n planar lines in $O(n^2)$ time and $O(n)$ space with a topological line and is a critical ingredient in several space and time efficient algorithms (e.g. [12], [6], [9], [7]). The technique has been adapted for specific applications (e.g. [10]) and has a

*** Partially supported by NSF grant EIA-99-96237
[†] Partially supported by NSF RUI grant 9731804

D. Mount and C. Stein (Eds.): ALENEX 2002, LNCS 2409, pp. 155–165, 2002.
© Springer-Verlag Berlin Heidelberg 2002

useful variant ([1]). The algorithm and its variations have been implemented by several groups (e.g. [11], [9]...).

Computational Geometry libraries such as LEDA and CGAL offer implementations of related but slightly less efficient line sweep algorithms. To the best of our knowledge, no robust code dealing with all degeneracies is currently available.

In contrast, the method proposed here is simple to compute and does not require special preprocessing. The modified algorithm was coded and the code was verified on different types and sizes of data sets. The code was incorporated in an implementation of an algorithm for statistical data analysis.

2   The Topological Sweep Algorithm [5]

Let H be a arrangement of n lines in the plane. Vertical line sweep could report all intersection pairs sorted in order of x-coordinate in $\Theta(n^2 \log n)$ time and $O(n)$ space (e.g. [3], [4]). If one only needs to report the intersection points of the lines according to a partial order related to the levels in the arrangement, greater efficiency is possible. To report all the intersection points of the lines, in quadratic time and linear space, we use a topological line (*cut*) that sweeps the arrangement. A topological line is a monotonic line in y-direction, which intersects each of the n lines in the arrangement exactly once. The cut is specified by the sequence of edges, one per line, each intersected by the topological line. A sweep is implemented by starting with the leftmost cut which includes all semi-infinite edges ending at -∞, and pushing it to the right until it becomes the rightmost cut, in a series of elementary steps. An elementary step is performed when the topological line sweeps past a vertex of the arrangement. To keep the sweep line a topological line we can only sweep past a vertex which is the intersection point of two consecutive edges in the current cut (a *ready* vertex). See Fig. 1 for an example of an arrangement of 7 lines and a topological cut.

2.1   Data Structures

The algorithm uses the following data structures:

- E[1:n] is the array of line equations. E[i] $= (a_i, b_i)$ if the ith line of the arrangement sorted by slope a_i is $y = a_i x + b_i$. This array is static.
- HTU[1:n] is an array representing the upper horizon tree. HTU[i] is a pair (λ_i, ρ_i) of indices indicating the current lines that delimit the segment of l_i in the upper horizon tree to the left and the right respectively. If the segment is the leftmost on l_i then $\lambda_i = -1$. If it is the right most then $\rho_i = 0$.
- HTL[1:n] represents the lower horizon tree and is defined similarly.
- I is a set of integers represented as a stack that correspond to points that are currently *ready* to be processed. If i is in I then c_i and c_{i+1} (the ith and $i^{th} + 1$ lines of the cut) share a common right endpoint and represent a legal next move for the topological line.

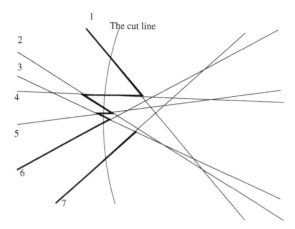

Fig. 1. An arrangement of 7 lines and a cut defined by its edges. The cut edges are marked as bold lines

- M[1:n] is an array holding the current sequence of indices from $E[i]$ that form the lines m_1, m_2..m_n of the cut.
- N[1:n] is a list of pairs of indices indicating the lines delimiting each edge of the cut, one from the left and the other from the right.

2.2   Horizon Trees and the Construction of the Cut

The upper (respectively lower) horizon tree of the cut is constructed by extending the cut edges to the right. When two edges intersect only the one of higher (respectively lower) slope continues to the right (see Fig. 2 for an example of upper and lower horizon trees). The upper (lower) horizon tree is initially created by inserting the lines in decreasing (increasing) order of slope into the structure. To insert line l_{k+1} we begin at its endpoint on the left boundary. We walk in counterclockwise order around the bay formed by the previous lines to find the intersection point of l_{k+1} with an edge. Given the lower and upper horizon trees, the right endpoint of the cut is identified by the leftmost of the two right delimiters of HTU[i] and HTL[i].

Time Complexity. Each elementary step and the accompanying updates to the upper and lower horizon trees, the cut arrays, and the ready stack take amortized constant time. There are at most n^2 elementary steps and therefore the total time complexity is $O(n^2)$. The original paper [5] contains a detailed analysis.

2.3   Dealing with Degeneracies

The original paper proposes dealing with degenerate cases such as parallel lines or multiple concurrent lines by using a primitive procedure to treat two parallel

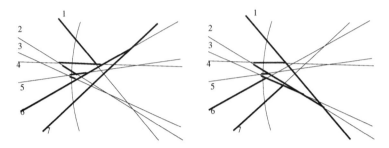

Fig. 2. Upper and Lower Horizon Trees

Table 1. Upper and Lower Horizon Trees

	Upper Horizon tree		Lower Horizon tree	
Line	Left delimiter	Right delimiter	Left delimiter	Right delimiter
1	-∞	4	-∞	∞
2	4	5	4	1
3	5	6	5	2
4	2	6	2	1
5	3	6	3	2
6	-∞	7	-∞	3
7	-∞	∞	-∞	3

lines as non-parallel and three concurrent lines as non-concurrent ([5], [4]). This original method generates zero length edges and vertices at infinity and then performs topological sweep on an arrangement in general position that is ϵ different from the original arrangement. The original method demands the computation of the power series expansion (in the perturbation parameter ϵ) of the value of a determinant until the first non-zero term is encountered (see also Gomez *et al* [8]). Edelsbrunner and Guibas also propose to detect degeneracies in configurations in $O(n^2)$ time and $O(n)$ space by checking for edges of length zero.

3   The Modified Algorithm

3.1   Data Structures

The modified algorithm uses many of the same data structures as in the original algorithm: $E[]$ holds the line equations; $M[]$, the order of the lines along the cut; $HTL[]$ and $HTU[]$, the horizon trees. The cut data structure $N[]$ and the entries in the ready stack I, however, each have an additional field. $N[i]$ is now a triplet $(\lambda_i, r_{up,i}, r_{down,i})$ of indices. If the right endpoint of $N[i]$ is generated by its intersection with a line from above (resp. from below), then $r_{up,i}$ (resp. $r_{down,i}$) is the index of that line; otherwise $r_{up,i}$ (resp. $r_{down,i}$) is null. At least one is non-null, but both may be in instances where the right endpoint of the cut edge is the intersection point of three or more lines. The left delimiter λ_i can be any line that intersects the current edge at its left endpoint. Given the lower and upper horizon trees, these delimiters are computed in constant time.

Each entry in the augmented stack I of ready vertices is now a pair of integers (i, k) where the segments of the lines m_i, m_{i+1}... m_{i+k} (the ith until $(i + k)^{th}$ edges of the cut) share a common right endpoint and represent a legal next move. These modifications do not change the asymptotic space complexity of the algorithm.

One new data structure is needed. $MATCH[i]$ is a pair of indices pointing to the uppermost and lowermost cut edges that currently share the same right end point as line l_i. $MATCH[i]$ is initialized at the beginning of the algorithm and reset to the pair (i, i) every time l_i participates in an elementary step, meaning that the match to line l_i is trivially line l_i itself. $MATCH$ is updated at the conclusion of each elementary step to detect new alignment of right end points. The fact that the edges along the cut sharing the same right endpoint form at most one connected component of adjacent edges (see Lemma 1 below) allows us to update only two boundary edges (top and bottom of the edges found so far) and ignore the intervening entries.

3.2   Computing Ready Vertices

Define a *matching pair* as a pair of consecutive lines l_i, l_j in the cut where $r_{up,i}$ is j and $r_{down,j}$ is i. When at most two lines participate in an intersection, a *matching pair* implies a *ready vertex*. For the more general case, we define a *matching sequence* of consecutive lines $l_i, ..., l_j$ in the cut where every adjacent pair of lines forms a *matching pair*. A *ready vertex* is generated by a *complete matching sequence* where the bottom line is l_i and the top line is l_j and in which $r_{down,i}$ and $r_{up,j}$ are both *null*.

After each update to the cut, we test whether two newly adjacent lines form a matching pair. If so, this new pair either augments a *matching sequence* already represented in $MATCH[]$ or initializes a new one. Updating $MATCH[]$ and checking whether the *matching sequence* is not *complete* takes constant time (see Lemma 3 below).

3.3   Parallel Lines and Identical Lines

Parallel lines create an intersection point at infinity. Their intersection point is not treated differently than that of any other line. Identical lines are lines that have the same slope and y-intercept. We currently treat identical lines together as a single line, under the assumption that the application that calls topological sweep will note and handle the impact of duplicate lines. For example, our application code, that computes depth contours based on the levels of the arrangement, handles this phenomenon (see [9]).

3.4   Additional Changes to the Algorithm

Tests called 'above' and 'closer' are used when the data structures are constructed and updated. Each comparison has three possible outcomes (instead of two): TRUE, FALSE and EQUAL. The EQUAL state is not part of the original algorithm since it is only generated in degenerate cases. (see 6). To deal with parallel lines, special test cases were added, that check if the lines in question are parallel or not. The tests are peformed during the initialization phase of the algorithm

Each update step may involve two or more updates, depending on the size of the intersection. The different data structures demand different update strategies. One update method is to replace each of the lines from i to j in a matching sequence. Lines are paired from the outermost lines to the innermost (i to j, $i+1$ to $j-1$, etc). This procedure is used to update the left delimiters of the cut and the upper and lower horizon trees and to update the order of the lines along the cut (M). Another method of updating is by computing the updated lines one by one. This method is used to update the right delimiters of the horizon trees and the cut. The horizon trees must be processed consecutively, otherwise the cut will not be updated correctly.

4   Example of a Degenerate Case

For the arrangement of 5 lines depicted in Fig. 3, the information in Table 4, describing the current cut, is computed from the upper and lower horizon trees. For each line l_i, the right delimiter of the lower horizon tree initially becomes $r_{up,i}$ and the right delimiter of the upper horizon tree initially becomes $r_{down,i}$. In the next step each pair of delimiters is compared to see which implied intersection point is *closer*: the delimiters for line 1 are ∞ and 5, therefore only 5 remains and the right-up delimiter of line 1 becomes *null*. Line 2 and 5 are similar to line 1 since one horizon tree delimiter is ∞ and it becomes *null*. The delimiters of lines 3 and 4 intersect in the same point and hence both remain.

In the *matching procedure* we only look for matches for the edges that were created after processing the last ready point (the intersection between lines 1 and 4). We start by looking for a match for line 1. Since right-down of line 1 is 5 and right-up of line 5 is 1, we have a matching pair. Since both the other delimiters of lines 1 and 5 are *null*, we will have a *complete matching sequence* and must update the stack I with the pair $(4, 1)$. We now consider line 4. Since right-up of line 4 is 3 and right-down of line 3 is 4 we have a match. We continue to look for an existing matching sequence with line 3. Our $MATCH$ data structure indicates a matching sequence starting with line 3 and ending with line 2 (consisting of just one pair where right-down of line 2 is 3 and right-up of line 3 is 2) which we expand to a matching sequence starting at line 4 and ending with line 2: $MATCH[4]$ and $MATCH[2]$ are both the pair $(2, 4)$. Note that we do not bother to update $MATCH[3]$. The right-up delimiter of 2 is null so we do not need to check for additional matches above. But the right-down of line 4 is still unmatched as right-down of 4 is 5 and right-up of line 5 is *null*. The matching sequence remains *incomplete*.

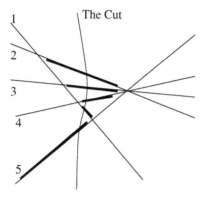

Fig. 3. Example of a degenerate case

Line i	λ_i	$r_{\text{up},i}$	$r_{\text{down},i}$
1	4	*null*	5
2	1	*null*	3
3	1	2	4
4	1	3	5
5	$-\infty$	1	*null*

5 Performance Analysis

Lemma 1. *The set of edges along any cut that contain the same intersection point as their right endpoint form at most one connected component.*

Proof. Assume the edges that contain p as their right endpoint form more than one connected component (Fig. 4). Let the edges be $p_1, p_2, \ldots$ and the associated lines be $l_{p_1}, l_{p_2}, \ldots$. Since there is more than one connected component there exist at least one line l_k between p_i and p_{i+1} that contains p but whose current cut edge c_k does not have p as a right endpoint. Assume the right end-point q of c_k is delimited by a line l. If the slope of l is smaller (resp. larger) than that of l_k, then l intersects l_i (resp. l_{i+1}) at a point r between points p and q. Since point q has not yet been processed, point r is not yet ready and has not been processed, and therefore the edge p_i (resp. p_{i+1}) cannot yet be part of the cut. Contradiction.   □

Lemma 2. *The total cost of updating HTU (HTL) through all the elementary steps is $O(n^2)$.* See original paper [5].

Lemma 3. *The total cost of comparing adjacent edges (computing the ready points) through all the elementary steps is $O(n^2)$*

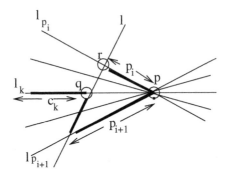

Fig. 4. Proof of Lemma 1

Proof. Each edge, c, that terminates at an intersection can generate at most 3 comparison tests. First a matching test is performed with the edges above and below. If a matching edge m is found, the $MATCH[m]$ will also be tested. According to Lemma 1 at most one of the edges that delimits c above and below can have a MATCH, otherwise more than one connected component exists. Therefore at most 3 tests will be generated. There is no need to perform additional tests since if another match exists it should have been found earlier, when the edges that form it were investigated. The total number of edges that enter all the intersections is n^2 and therefore the total complexity is at most $O(n^2)$. □

Lemma 4. *The time complexity of the algorithm is $O(n^2)$.*

Proof. Initialization of all data structures takes linear time after $O(n \log n)$ time to sort the lines by slope. Each elementary step takes an amortized constant time, as shown in Lemmas 2 and 3. There are $O(n^2)$ elementary steps. The total time complexity is therefore $O(n^2)$. □

6   Discussion

To ensure the correctness of the method, we verify that rounding errors do not produce irrecoverable mistakes.

One test compares slopes to compute whether two lines are parallel. The answer (*parallel* or *not*) will be as good as the data set provided: no round-off errors occur since there is no computation involved.

Other tests compare computed values of x or y coordinate to check if a point / intersection of the arrangement is *above* or *closer* than another point / intersection and return one of *true*, *false* or *equal*. These computations involve multiplications of at most two values. When the input values are floats, using double precision to compare the computed values ensures that only errors that return an *equal* answer instead of *true* or *false* can occur. No answer can have the opposite value to the real value (e.g. *true* instead of *false* or *true/false* instead of *equal*). When receiving an answer that is *equal* instead of *not-equal*, three or more lines of the arrangement will be treated as passing through the same point although they do not. Effectively a triangular face is contracted to a point but otherwise there is no change in the topological structure of the arrangement.

7   Experimental Results

7.1   Analysis Method

Our experiments checked the behavior of the code in simple and in degenerate cases. We created sets of lines by generating n random numbers representing the slopes of the lines and n random numbers representing the y-intercept of the lines. We paired the numbers to receive a representation of an arrangement of n lines.

At the end of a correct sweep all the right delimiters must be ∞. If a mistake occurs the new topology will not allow the sweep to continue until the rightmost point and will stop too early, not reaching ∞. Hence, a check that all the right delimiters are ∞ is made to verify that the sweep was performed correctly.

To generate parallel lines we chose m numbers out of the n slopes generated above and m numbers out of the n y-intercepts generated above. We paired these at random where m is 5 percent of n, and used the $n+m$ pairs as our data set. This ensured that at least m pairs have the same slopes and are therefore parallel or coincident.

To generate multiple concurrent lines we computed the n^2 intersection points of the original arrangement we generated. We than randomly chose n points out of the n^2 intersections and used their dual lines as our data set. The dual line of point (a, b) is the line $y = ax + b$. If more than two points are on the same line their dual lines share a common point and we get a degenerate case of multiple concurrent lines. By selecting n intersection points from the n^2 intersections of the original set and taking their dual, the probability of choosing more than one point on the same line, hence creating an intersection of multiple lines in the dual, is relatively high.

7.2   Results

Our code is written in C++, does not use any geometric libraries for computations, but uses GEOMVIEW for visualization of the output. The code is built modularly and can be easily modified. It is located at:

http://www.eecs.tufts.edu/r/geometry/sweep.

It was tested on a Sun Microsystems Ultra 250 SPARC processor, 400 MHz, and was compiled using the GNU C++ compiler. We ran our code on 10 different data sets of each size and type that were generated as described above. The average results are presented in Fig. 5.

The implementation is not assumed to perform better than other topological sweep implementations in terms of running time. Instead it treats cases that

other implementations were slow or unable to handle. Our package creates data
that is used for display purposes. If display is not needed the computation can
be streamlined further.

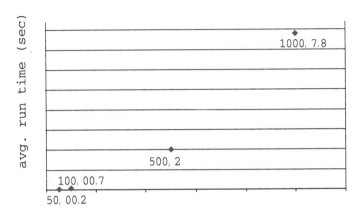

Fig. 5. Measured execution times (in seconds) for different set types

8   Future Research

8.1   Guided Topological Sweep

Guided topological sweep is a topological sweep that maintains additional order
criteria without penalizing performance. For example, it is possible to guarantee
that whenever a vertex is swept, the edges that are k lines below it and k
lines above it are aligned with it without changing asymptotic complexity. This
method is used for LMS regression in $O(n^2)$ time (see [6]). We plan to expand
our algorithm to enable this type of sweep.

8.2   Topological Sweep in Higher Dimensions

Most statistical (and other) data sets are multi-dimensional. There are some theoretical algorithms for high dimensions but few of them have been implemented.
We are working to expand our implementation to higher dimensions and use this
as a sub-procedure for high-dimensional applications.

8.3   Applications

Degenerate data sets that include more than one point with the same x-
coordinate or concurrent points form a large part of the available (and interesting

to investigate) sets. An earlier implementation of the depth contours algorithm ([9]) has been reused and expanded. By calling the new topological sweep procedure it can now handle degenerate data sets.

9   Conclusion

We present an efficient application of the topological sweep algorithm that uses extended data structures instead of numerical methods to deal with degenerate case. The new data structures use less than 1.5 times as much space as the original data structures.

References

[1] Te. Asano, Leonidas J. Guibas, and T. Tokuyama. Walking on an arrangement topologically. *Internat. J. Comput. Geom. Appl.*, 4:123–151, 1994.

[2] Christoph Burnikel, Kurt Mehlhorn, and Stefan Schirra. On degeneracy in geometric computations. In Daniel D. Sleator, editor, *Proceedings of the 5th Annual ACM-SIAM Symposium on Discrete Algorithms*, pages 16–23, Arlington, VA, January 1994. ACM Press.

[3] Mark de Berg, Mark van Kreveld, Mark Overmars, and Otfried Schwarzkopf. *Computational Geometry Algorithms and Applications.* Springer-Verlag, Berlin Heidelberg, 1997.

[4] H. Edelsbrunner. *Algorithms in Combinatorial Geometry*, volume 10 of *EATCS Monographs on Theoretical Computer Science.* Springer-Verlag, Heidelberg, West Germany, 1987.

[5] H. Edelsbrunner and Leonidas J. Guibas. Topologically sweeping an arrangement. *J. Comput. Syst. Sci.*, 38:165–194, 1989. Corrigendum in 42 (1991), 249–251.

[6] H. Edelsbrunner and D. L. Souvaine. Computing median-of-squares regression lines and guided topological sweep. *J. Amer. Statist. Assoc.*, 85:115–119, 1990.

[7] J. Gil, W. Steiger, and A. Wigderson. Geometric medians. *Discrete Mathematics*, 108:37–51, 1992.

[8] F. Gomez, S. Ramaswami, and G. Toussaint. On removing non-degeneracy assumptions in computational geometry. In *Algorithms and Complexity (Proc. CIAC' 97)*, volume 1203 of *Lecture Notes Comput. Sci.*, pages 86–99. Springer-Verlag, 1997.

[9] K. Miller, S. Ramaswami, P. Rousseeuw, T. Sellares, D. Souvaine, I. Streinu, and A. Struyf. Fast implementation of depth contours using topological sweep. In *Proceedings of the Twelfth ACM-SIAM Symposium on Discrete Algorithms*, pages 690–699, Washington, DC, January 2001.

[10] M. Pocchiola and G. Vegter. Topologically sweeping visibility complexes via pseudo-triangulations. *Discrete Comput. Geom.*, 16:419–453, December 1996.

[11] H. Rosenberger. Order k Voronoi diagrams of sites with additive weights in the plane. M.Sc. thesis, Dept. Comput. Sci., Univ. Illinois, Urbana, IL, 1988. Report UIUCDCS-R-88-1431.

[12] Emo Welzl. Constructing the visibility graph for n line segments in $O(n^2)$ time. *Inform. Process. Lett.*, 20:167–171, 1985.

Acceleration of K-Means and Related Clustering Algorithms

Steven J. Phillips

AT&T Labs-Research
180 Park Avenue, Florham Park, NJ 07932
phillips@research.att.com

Abstract. This paper describes two simple modification of K-means and related algorithms for clustering, that improve the running time without changing the output. The two resulting algorithms are called COMPARE-MEANS and SORT-MEANS. The time for an iteration of K-means is reduced from $O(ndk)$, where n is the number of data points, k the number of clusters and d the dimension, to $O(nd\gamma + k^2d + k^2\log k)$ for SORT-MEANS. Here $\gamma \leq k$ is the average over all points p of the number of means that are no more than twice as far as p is from the mean p was assigned to in the previous iteration. COMPARE-MEANS performs a similar number of distance calculations as SORT-MEANS, and is faster when the number of means is very large. Both modifications are extremely simple, and could easily be added to existing clustering implementations.

We investigate the empirical performance of the algorithms on three datasets drawn from practical applications. As a primary test case, we use the Isodata variant of K-means on a sample of 2.3 million 6-dimensional points drawn from a Landsat-7 satellite image. For this dataset, γ quickly drops to less than $\log_2 k$, and the running time decreases accordingly. For example, a run with $k = 100$ drops from an hour and a half to sixteen minutes for COMPARE-MEANS and six and a half minutes for SORT-MEANS. Further experiments show similar improvements on datasets derived from a forestry application and from the analysis of BGP updates in an IP network.

1 The K-Means Algorithm

Let P be a set of n points in R^d, and let k be a non-negative integer. Suppose we wish to cluster the points of P, partitioning them into k classes $C_1, \ldots, C_k$ so that each class consists of points that are close together. K-means [7] is a widely used algorithm for clustering, and sees frequent application in such diverse fields as astronomy, remote sensing, and speech recognition.

The basic operation of K-means is the iteration of the following two steps. Let $\mu_1, \ldots, \mu_k$ be the means of the classes found in the previous iteration.

1. Assign each point $p \in P$ to the class C_j that minimizes $d(p, \mu_j)$.
2. Recalculate the means: for each $j \in \{1 \ldots k\}$, set μ_j to be the mean of the points assigned to C_j in Step 1.

D. Mount and C. Stein (Eds.): ALENEX 2002, LNCS 2409, pp. 166–177, 2002.
© Springer-Verlag Berlin Heidelberg 2002

K-means can be started with any set of initial means, and ends when no point changes class in Step 1. It is easy to show that the distortion $(\sum_j \sum_{p \in C_j} d(p, \mu_j)^2)$ is monotonically decreasing, and that K-means ends with a local minimum of the distortion.

1.1   The Isodata Algorithm

There are a number of heuristics for helping K-means avoid bad local minima, such as when some means get "orphaned", having no points assigned to them. Because much of the data used in this paper is satellite imagery, we use a heuristic that is extensively used in such remote sensing applications, called Isodata (Iterative Self-Organizing Data Analysis Technique) [10,6].

Isodata specifies an initial set of means and a set of rules for discarding, splitting, and merging classes. In each dimension i, let μ_i be the mean and σ_i the standard deviation of the points. The initial means are spaced at even intervals on the line between $(\mu_1 - \sigma_1, \ldots, \mu_k - \sigma_k)$ and $(\mu_1 + \sigma_1, \ldots, \mu_k + \sigma_k)$. When a class size falls below `minimum_class_size` the class is discarded. If the number of classes is less than k, and some class has standard deviation higher than `maximum_standard_deviation` in some dimension i, then the class is split in two, with the new means a standard deviation apart in dimension i. There are rules for tie-breaking (when more than one class has high standard deviation), and the parameter `minimum_distance` determines when nearby classes should be merged. For further details, see Jensen [6].

1.2   Contributions of This Paper

The ideas described here reduce the running time of K-means, without changing the output. The resulting algorithms are COMPARE-MEANS, which takes advantage of the clustering obtained in each iteration to avoid making unnecessary comparisons between data points and cluster means in the next iteration, and SORT-MEANS, which makes a similar number of comparisons, while improving the running time further by reducing overhead.

The degree of acceleration improves with the quality of the clustering; this is quantified in Section 2.3. For our algorithms to be useful, not only must the data set have clusters, but K-means must also be finding a reasonable clustering. We therefore chose to study data sets drawn from applications where K-means (or the Isodata variant) see frequent and successful application. The results of the study are contained in Section 4. We chose not to do experiments on synthetic data, as there is a huge range of data distributions that could be used, and it is not clear which, if any, would accurately model data that would be observed in practice.

2   Accelerating K-Means and Isodata

2.1   Acceleration by Comparing Means

In Step 1 of the K-means algorithm, each point p is compared against each mean μ_j, resulting in $O(nk)$ comparisons and a running time of $O(ndk)$ per iteration. The algorithm COMPARE-MEANS uses a simple approach to avoid many comparisons that are provably unnecessary.

Consider a point $p \in P$ and consider two means μ_i and μ_j. Using the triangle inequality, we have $d(\mu_i, \mu_j) \leq d(p, \mu_i) + d(p, \mu_j)$, so

$$d(p, \mu_j) \geq d(\mu_i, \mu_j) - d(p, \mu_i).$$

Therefore if we know that $d(\mu_i, \mu_j) \geq 2d(p, \mu_i)$ we can conclude that $d(p, \mu_j) \geq d(p, \mu_i)$ without having to calculate $d(p, \mu_j)$.

Algorithm COMPARE-MEANS simply precomputes the distance $d(\mu_i, \mu_j)$ for each pair of means before each iteration. Then before comparing a point p to a mean μ_j, we perform the above test using the closest known mean to p. Each iteration is as follows:

1a. Calculate all inter-mean distances $D[i][j] = d(\mu_i, \mu_j)$.
1b. For each point $p \in P$, let c be the class p was assigned to in the previous iteration (by default $c = 1$ for the first iteration), and do the following steps. At the end, p is assigned to class minClass.

```
minDist = d(p, mu[c]);
minClass = c;
for (i=1; i<=k; i++) {
    if (D[i][minClass]) >= 2*minDist || i==c)
        continue;
    dist = d(p, mu[i]);
    if (dist<minDist) {
        minDist = dist;
        minClass = i;
    }
}
```

Note that we initialize minDist using the mean p was assigned to in the previous iteration. The is because for many datasets, in all but the first few iterations most points don't change class, so the last assigned mean is frequently the closest mean.

2.2   Acceleration by Sorting Means

A further speedup is obtained by first sorting the means in order of increasing distance from each mean. An iteration of SORT-MEANS is as follows.

1a. Calculate all inter-mean distances $D[i][j] = d(\mu_i, \mu_j)$, then construct the $k \times k$ array M, in which row i is a permutation of $1 \ldots k$, representing the classes in increasing order of distance of their means from μ_i.

1b. For each point $p \in P$, let c be the class p was assigned to in the previous iteration, and do the following steps. At the end, p is assigned to class minClass.

```
inClassDist = d(p, mu[c]);
minDist = inClassDist;
minClass = c;
for (i=2; i<=k; i++) {
    theClass = M[c][i];
    if (D[c][theClass]) >= 2*inClassDist
        break;
    dist = d(p, mu[theClass]);
    if (dist<minDist) {
        minDist = dist;
        minClass = theClass;
    }
}
```

In words, we compare point p against the means in increasing order of distance from the mean μ_c that p was assigned to in the previous iteration. If we reach a mean that is far enough from μ_c, we know we can skip all the remaining means and continue on to the next point. In this way, SORT-MEANS avoids the overhead of looping through all the means.

Note that we could instead use the expression (inClassDist + minDist) in place of 2*inClassDist, which would reduce the number of comparisons for some points that change class during the iteration. We chose not to do this in the implementation described in this paper, as the simpler expression allows us to easily avoid taking square roots in the distance calculations (recording squared distances in D[][], inClassDist and minDist).

For the first iteration of SORT-MEANS we may have an initial assignment of points to means (see for example the discussion of bootstrapping at the end of Section 4.1). If not, we can arbitrarily use $c = 1$ for all points, and unless otherwise noted, this is done for simplicity in the experiments below. A slight reduction in the number of comparisons can be obtained by instead using COMPARE-MEANS for the first iteration; the reduction is significant when we have a good initial set of means, but we don't have a good initial assignment of points to those means.

2.3 Analysis

In an iteration of K-means, let γ be the average over all points p of the number of means that are no more than twice as far as p is from the mean p was assigned to in the previous iteration. The number of comparisons that SORT-MEANS makes in the iteration is at most $n\gamma$, rather than nk. Note that $\gamma \leq k$. One can construct datasets for which γ is close to k, however for datasets on which K-means finds a good clustering, with most points being much closer to their assigned mean than to most other means, we can expect γ to be much smaller than k. The running time of an iteration is $O(nd\gamma + k^2d + k^2 \log k)$, where the second and third terms are the time taken to compare and sort the means.

The number of comparisons made by COMPARE-MEANS is harder to characterize. If a point p does not change class in the iteration, then COMPARE-MEANS and SORT-MEANS compare p to the same set of means. For points that do change class, we expect COMPARE-MEANS to generally perform slightly fewer comparisons (because `minDist` < `inClassDist`), and indeed this is true for all the instances described below. The overhead of COMPARE-MEANS (time not spent comparing points to means) is $\Theta(k^2d + nkd)$ which is smaller than the overhead for SORT-MEANS when $nd = o(k \log k)$, i.e. when the number of means is very large.

3   Related Work

The running time per iteration of the (unaccelerated) K-means is $O(nkd)$ per iteration. The present paper gives a method for reducing the dependence of the running time on k. The previous work on accelerating K-means is a sequence of papers by Moore et al. [8,9], which develop the following idea for reducing the dependence on n. First, build a data structure such as a k-d-tree or metric tree, which partitions the dataset and essentially gives a pre-classification of the data. Then, for each iteration of K-means, rather than running through the point set, one can traverse the tree. At each node in the tree, it may be possible to show that some classes cannot contain any of the points represented by that node. When the search reaches a node for which there is only one class that provably contains all the points represented by the node, the points can be assigned to that class as a group. The assignment is fast as long as "cached sufficient statistics" (such as the mean of the represented points) are kept at each node.

Section 4.3 analyzes a dataset used by Moore [8], and finds that SORT-MEANS performs about the same number of comparisons. However, the acceleration described in this paper has two advantages: its extreme simplicity and low overhead. Regarding simplicity, the loop of Step 1b of SORT-MEANS differs from the innermost loop of standard K-means by having just two extra array accesses and a conditional termination of the loop, while COMPARE-MEANS is simpler still. The acceleration can be very easily added to existing or future implementations of K-means. The low overhead is also important, especially on lower-dimensional data — on the Landsat data described below, SORT-MEANS performs more comparisons than COMPARE-MEANS, but runs much faster because little time is spent outside of comparisons.

Because the approach of Moore et al. reduces dependence on n, while our approach reduces dependence on k, it is an interesting open question whether the two approaches can be combined to achieve further acceleration.

While this paper considers only K-means and related algorithms, there is a large body of work on other algorithms for clustering multi-dimensional data; see for example [2,3,4].

4   Experimental Results

The algorithms were run on a collection of datasets derived from a Landsat satellite image, a forestry application, and an Internet application. The behaviour of COMPARE-MEANS and SORT-MEANS are investigated in detail on the Landsat image, while the other datasets are used to verify the effectiveness of the acceleration in very different situations.

4.1   Landsat Dataset

The primary dataset used here is drawn from a Landsat 7 image of the north of Vietnam, specifically the image in path 128, row 44, taken on December 27 1999. The image has three visible and three near-infrared to mid-infrared bands. Each band is a 7051×7911 matrix of brightness values, each a number in the range 0 to 255. A sample of 2,370,686 pixels was drawn by selecting every fourth row and every fourth column of the image, retaining only pixels with non-zeroes in all bands.

In all runs described below, the influence of using Isodata remained limited to the first few iterations. Some classes were discarded and others split during early iterations; no discarding or splitting occurs at all for $k < 330$ (so the behaviour is identical to K-means), and for $k < 400$, all discarding and splitting occurs in the first two iterations. The Isodata parameters determining this behaviour were as follows: `minimum_class_size` 10 points, `maximum_standard_deviation` 10.0, `minimum_distance` 0 (so no classes get merged).

The accelerated algorithms were run with k ranging from 10 to 500 in increments of 10, and the standard algorithm was run with 10, 50, 100, 200, 300, 400 and 500 means. All runs were terminated after 100 iterations, although none had converged (i.e. reached an iteration in which no point changed class).

Figure 1 investigates the quantity γ, the number of comparisons per point made by SORT-MEANS (the number of times it computes the distance between a point and a mean) in each iteration. Considering the line for $k = 100$, we see that in the first iteration, a typical point is compared against 82 means, out of a maximum of 100. However, the number of comparisons per point drops precipitously, reaches 6 in the 5th iteration, and is below 5.1 by the 15th iteration. It quickly levels out, dropping only to 4.95 by the 100th iteration. For comparison, the number of comparisons per point without acceleration is exactly k. Thus for all but the first few iterations in the run with $k = 100$, acceleration reduces the number of comparisons by a factor of 20. COMPARE-MEANS performs almost exactly as many comparisons as SORT-MEANS on this data.

Figure 2 shows that the reduction in the number of comparisons causes a corresponding reduction in the total running time (note the log scale). For 100

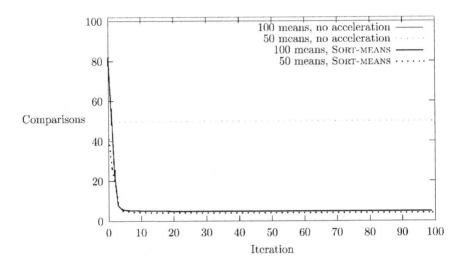

Fig. 1. Average number of comparisons per point (γ) in each iteration

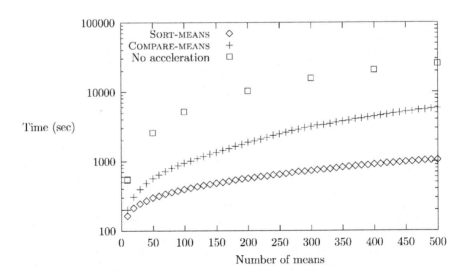

Fig. 2. Total runtime for varying numbers of means

means, the running time drops from 5115 to 945 seconds for COMPARE-MEANS and 390 seconds for SORT-MEANS, i.e. from an hour and a half to six and a half minutes. Quoted running times are for a Java implementation, running with the Java 1.3 runtime on a 850 MHz Pentium III running Windows 98. The same code was run on a Silicon Graphics Power Challenge with 270MHz MIPS R12000 IP27 processors, with running times consistently 4 to 4.5 times slower than those quoted here.

The overhead for comparing and sorting the means during these runs is very small. It is greatest for the run with 500 means, for which it is less than 400 milliseconds per iteration for SORT-MEANS, or about 4% of the total running time (and much less for COMPARE-MEANS).

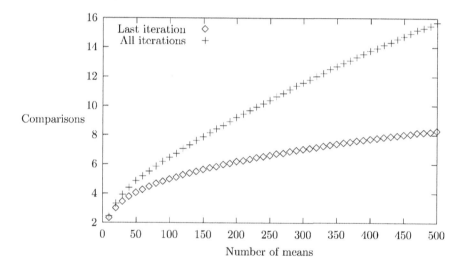

Fig. 3. Average number of comparisons per point averaged over all iterations, and in last iteration. Algorithm SORT-MEANS.

In Figure 3 we investigate how the average number of comparisons per point varies with the number of means for SORT-MEANS (COMPARE-MEANS is almost identical for this dataset). The upper line shows the number of comparisons per point averaged over all 100 iterations. This number is strongly influenced by the comparisons made in the first few iterations. The lower line shows only the number of comparisons per point in the last iteration. The latter value is of interest if we wish to run more than 100 iterations, and therefore get a better clustering of the points, as the number of comparisons per iteration is nearly monotonically decreasing. For this dataset, the number of comparisons per point in the last iteration is less than $\log_2 k$.

The difference between the upper and lower lines in Figure 3 can be reduced by using a better initial set of means, and by making a good initial assignment of points to classes. For example, consider the following pre-classification process for a 6-dimensional dataset. First, run 20 iterations of Isodata on a subsample of a tenth of the points; this gives a good initial set of means. Then generate an initial assignment of points to classes as follows: create a six-dimensional array *initial*, where $initial[i_0][i_1][i_2][i_3][i_4][i_5]$ is the closest mean to a hypothetic point whose value in dimension j is $-1.25+i_j/2$ standard deviations from the mean, for $i_j = 0 \ldots 5$. For each dimension j, generate an array dev_j that gives the number of standard deviations from the mean for each of the 256 possible pixel values. Then the initial assignment of each point is simply determined by indexing into the *dev* and *initial* arrays. For $k = 400$, this procedure reduces the time for the first iteration of SORT-MEANS on the large point set from 195 seconds to 36.7, the second iteration from 147 seconds to 5.2 and the total running time from 896 seconds to 581 seconds (including pre-classification). Because of the better initial set of means, the distortion at the end of 100 iterations on the large point set is smaller than without pre-classification. In fact, after 239 seconds and 30 iterations on the large set, the distortion is lower than after 100 iterations without pre-classification.

We can simplify this bootstrapping process, with only a small time penalty, by not calculating the *dev* and *initial* arrays, instead using COMPARE-MEANS for the first iteration on the large point set to take advantage of the good initial means. The time for the first iteration is then 52.2 seconds, giving a total running time of 597 seconds.

Other methods could be used to obtain a good initial clustering, for example the algorithm of Gonzalez [5], which computes a 2-approximation for the problem of minimizing the maximum intra-cluster distance.

4.2   More Landsat Datasets

First the Landsat 7 image was sampled at various frequencies, producing varying numbers of points. For example, for the smallest dataset, every eighth row and column of the original image was retained. For each sampling frequency, 3 dimensional data corresponding only to the visible bands was considered, in addition to the 6 dimensional data that includes the visible, near and mid-infrared bands. Each run used $k = 100$ and 100 iterations. The results for SORT-MEANS are shown in Table 1, where "avg comparisons" refers to the number of comparisons per point averaged over all iterations, and "end comparisons" refers to the last iteration. We see that the absolute speedup ranges from a factor of 13 to a factor of 16, consistent with the primary dataset above.

4.3   Forest Cover Type

The next dataset is the Forest Cover Type dataset from the UCI Knowledge Discovery in Databases Archive [1]. This dataset, which was also used as a test instance by Moore [8], relates forest cover type for 30 x 30 meter cells (obtained

Table 1. Performance on various samples of the Landsat 7 image.

Points	Bands	Comparisons Avg	End	Time (seconds) SORT-MEANS	No acceleration	Speedup
2408250	3	4.765	3.581	176	2872	16
1053635	6	6.428	4.951	182	2316	13
1070341	3	4.753	3.565	79	1279	16
592674	6	6.437	4.947	99	1275	13
602070	3	4.745	3.578	48	720	15

from US Forest Service Region 2 Resource Information System data) to various variables, such as distance from roads, elevation and soil type. There are 581012 data points in 54 dimensions (though Moore gives the number of data points as 150000). Ten of the dimensions contain integer quantitative data, while the remaining 44 are binary, though all 54 are handled identically by K-means.

Table 2. Performance on the Forest Cover Type dataset.

	Comparisons (avg)	Comparisons (last iteration)	Time
No acceleration	100.0	100.0	3:09:00
COMPARE-MEANS	5.488	5.253	14:47
SORT-MEANS	6.149	5.297	13:00

Table 2 shows the result of running 100 iterations with $k = 100$, and initial means randomly selected from the set of points. We see that Algorithm COMPARE-MEANS reduces the number of comparisons by a factor of 18, and reduces the time by a factor of 13. Algorithm SORT-MEANS gives a slight improvement in speed, giving a factor of 15 speedup over ordinary K-means.

The number of comparisons made by both algorithms compares is slightly higher than is reported in Moore [8], where a reduction of the number of comparisons by a factor of 19 is reported. However, it is difficult to make a direct comparison, as the reported number of data points is different. In addition, run time is not reported in [8], and it is not clear whether the k-d-tree operations constitute a significant fraction of run time.

4.4 BGP Updates

The last dataset involves the Border Gateway Protocol (BGP) running on the global Internet. All BGP updates received from the global Internet by AT&T were observed for 29 days. For each IP address block (or routing table entry), the log of the number of updates for each of the 29 days was recorded. The resulting real-valued data consists of 140980 "points" in 29 dimensions.

A cluster of similar points corresponds to a set of IP addresses all generating similar numbers of BGP updates on the same days, perhaps all affected by a single fault in the network. Clustering the dataset may therefore help to identify the source and extent of network faults. The logarithm is used because it better models the intuitive notion of a similar number of updates.

This dataset has a small natural cluster size, so the clustering algorithms were run with a large number of means (5000). The initial means were randomly chosen, and the runs converged after 27 iterations. For this data set, SORT-MEANS was run with a first iteration of COMPARE-MEANS, taking advantage of the widely separated initial means (see the end of Section 2.2 for more discussion). The performance of the algorithms is shown in Figure 3.

Table 3. Performance on the BGP updates dataset.

	Comparisons (avg)	Comparison / sort time	Total time
No acceleration	5000.0	0	15:51:16
COMPARE-MEANS	365.153	21:57	1:24:08
SORT-MEANS	366.933	42:54	1:34:55

Algorithm COMPARE-MEANS achieves a speedup of 11.3, and performs a factor of 13.6 fewer comparisons than standard Isodata. Unlike the previous datasets, the algorithm SORT-MEANS is slower than COMPARE-MEANS, even with a first iteration using COMPARE-MEANS, because of the overhead of sorting the large number of means. Indeed, algorithm SORT-MEANS spends almost half of its time sorting the means.

5   Conclusions and Extensions

This paper has presented two modifications for accelerating K-means: avoiding making unnecessary comparisons between data points and means by comparing means to each other (algorithm COMPARE-MEANS), and avoiding overhead by sorting the means (algorithm SORT-MEANS). The modifications offer very significant speedups for K-means and related classification algorithms on a wide range of datasets. The modifications are simple enough to be easily added to existing implementations of clustering algorithms.

Algorithm SORT-MEANS is significantly faster than algorithm COMPARE-MEANS (except when the number of means is very large), and the difference is most pronounced when the number of dimensions of the dataset is small.

Acknowledgements. Thanks to the Center for Biodiversity and Conservation at the American Museum of Natural History for the opportunity to learn about remote sensing and for the use of the Landsat 7 image, to Ned Horning at Spatial

Support Services for conversations on remote sensing, and to Sanjoy Dasgupta and Howard Karloff at AT&T Labs-Research for conversations on the K-means algorithm and clustering. Thanks to Carsten Lund and Nick Reingold at AT&T Labs-Research for providing the BGP-update dataset.

References

1. S. D. Bay. The UCI KDD Archive [http://kdd.ics.uci.edu]. Irvine, CA: University of California, Department of Information and Computer Science., 1999.
2. P. S. Bradley, U. M. Fayyad, and C. Reina. Scaling clustering algorithms to large databases. In *Knowledge Discovery and Data Mining*, pages 9–15, 1998.
3. A. Dempster, N. Laird, and D. Rubin. Maximum-likelihood from incomplete data via the EM algorithm. *J. Royal Statistical Society B*, 39:1–38, 1977.
4. F. Farnstrom, J. Lewis, and C. Elkan. Scalability for clustering algorithms revisited. *SIGKDD Explorations*, 2(1):51–57, 2000.
5. T. F. Gonzalez. Clustering to minimize the maximum intercluster distance. *Theoretical Computer Science*, 38:293–306, 1985.
6. J. R. Jensen. *Introductory Digital Image Processing, A Remote Sensing Perspective.* Prentice Hall, Upper Saddle River, NJ, 1996.
7. J. MacQueen. Some methods for classification and analysis of multivariate observations. In *Proc. Fifth Berkeley Symposium on Mathematics, Statistics and Probability*, volume 1, pages 281–296, 1967.
8. A. W. Moore. The anchors hierarchy: Using the triangle inequality to survive high dimensional data. In *Proc. UAI-2000: The Sixteenth Conference on Uncertainty in Artificial Intelligence*, 2000.
9. D. Pelleg and A. W. Moore. Accelerating exact k-means algorithms with geometric reasoning. In *Proc. Fifth International Conference on Knowledge Discovery and Data Mining*. AAAI Press, 1999.
10. J. T. Tou and R. C. Gonzalez. *Pattern Recognition Principles*. Addison-Wesley, Reading, MA, 1977.

STAR-Tree: An Efficient Self-Adjusting Index for Moving Objects*

Cecilia M. Procopiuc[1], Pankaj K. Agarwal[2], and Sariel Har-Peled[3]

[1] AT&T Research Lab, Florham Park, NJ 07932
magda@cs.duke.edu
[2] Duke University, Durham, NC 27708-0129
pankaj@cs.duke.edu
[3] University of Illinois, Urbana, IL 61801
sariel@cs.uiuc.edu

Abstract. We present a new technique called STAR-tree, based on R*-tree, for indexing a set of moving points so that various queries, including range queries, time-slice queries, and nearest-neighbor queries, can be answered efficiently. A novel feature of the index is that it is self-adjusting in the sense that it re-organizes itself locally whenever its query performance deteriorates. The index provides tradeoffs between storage and query performance and between time spent in updating the index and in answering queries. We present detailed performance studies and compare our methods with the existing ones under a varying type of data sets and queries. Our experiments show that the index proposed here performs considerably better than the previously known ones.

1   Introduction

Motion is ubiquitous in the physical world. Several areas such as digital battlefields, air-traffic control, mobile communication, navigation system, geographic information systems, call for indexing moving objects so that various queries on them can be answered efficiently; see [13,11] and the references therein. The queries might relate either to the current configuration of objects or to a configuration in the future — in the latter case, we are asking to predict the behavior based on the current information. In the last few years there has been a flurry of activity on extending the capabilities of existing database systems to represent moving-object databases (MOD) and on indexing moving objects; see, e.g., [8, 12,13]. In this paper we propose a new indexing technique called *spatio-temporal self-adjusting R-tree*, referred to as STAR-tree for brevity, for indexing the trajectories of moving point objects so that range queries and their variants can be

* C.P. is supported by Army Research Office MURI grant DAAH04-96-1-0013 and an NSF grant CCR–9732787. P.A. is supported by Army Research Office MURI grant DAAH04-96-1-0013, by a Sloan fellowship, by NSF grants ITR–333–1050, EIA–9870724, EIA–997287, and CCR–9732787, and by a grant from the U.S.-Israeli Binational Science Foundation.

D. Mount and C. Stein (Eds.): ALENEX 2002, LNCS 2409, pp. 178–193, 2002.
© Springer-Verlag Berlin Heidelberg 2002

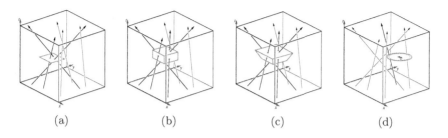

Fig. 1. Instances of Q1, Q2, Q2', and Q3 queries, respectively.

answered efficiently. Our technique is a natural extension of R*-tree [6] and is closely related to the index introduced by Šaltenis [11]. Unlike the earlier techniques (e.g. [10,11]), we neither reconstruct the index periodically nor do we require the index to be updated frequently. Instead, the index maintains certain auxiliary information and updates itself locally to ensure that queries are answered efficiently. The index provides tradeoffs between the storage used and the query performance and between the time spent in updating the index and answering queries.

Section 2 states our model and summarizes the related work and the main contributions of this paper. Section 3 describes the indexing scheme, the events at which the index is updated, and the query and update procedures. Section 4 discusses experimental results. We conclude in Section 5 by summarizing our work and suggesting further directions.

2   Our Contributions and Related Work

Data model. Let $S = \{p_1, \ldots, p_n\}$ be a set of n points in $\mathbb{R}^2$, each moving independently. The position of a point p_i at time t is given by $p_i(t) = (x_i(t), y_i(t))$. We use $\bar{p}_i = \bigcup_t (p_i(t), t)$ to denote the *graph* of the trajectory of p_i in the xyt-space. We assume the trajectories to be piecewise-linear functions. The user is allowed to change the trajectory of a point at any time.

Queries. The index we propose can be used to answer various types of both *range queries* and *nearest neighbor queries*, based on the current and future positions of the objects. We use the notation *now* to denote the current time. Specifically, we study the following three types of queries.

Q1. Given an axis-aligned rectangle R in the xy-plane and a time value $t_q \geq now$, report all points of S that lie inside R at time t_q, i.e., report $S(t_q) \cap R$; see Figure 1 (a).

Q2. Given a rectangle R and two time values $now \leq t_1 \leq t_2$, report all points of S that lie inside R at any time between t_1 and t_2, i.e., report $\bigcup_{t=t_1}^{t_2} (S(t) \cap R)$; see Figure 1 (b).

Q2'. Given two orthogonal rectangles R_1 and R_2 and two time values $now \leq$ $t_1 \leq t_2$, report all points of S that lie inside $R(t)$ at any time between $t_1 \leq t \leq t_2$, where $R(t) = [(t - t_1)R_2 + (t - t_1)R_1]/(t_2 - t_1)$. That is, report $\bigcup_{t=t_1}^{t_2} (S(t) \cap R(t))$; see Figure 1 (c).

Q3. Given a query point $q \in \mathbb{R}^2$ and a time value $t_q \geq now$, an ε-*approximate k-nearest neighbor query* requires reporting k points $p'_1, \ldots, p'_k \in S$ such that $d(q, p'_i(t_q)) \leq (1 + \varepsilon)d(q, p_i(t_q))$, $1 \leq i \leq k$, where $p_i \in S$, $1 \leq i \leq k$, and $p_1(t_q), \ldots, p_k(t_q)$ are the k nearest neighbors of q. In particular, for $\varepsilon = 0$ and $k = 1$ we obtain the nearest neighbor query over moving points; see Figure 1 (d).

Previous work. One direction of research has been to index the trajectories of points, either directly, or by mapping them to higher dimensional points [10,14]. This is not very efficient since trajectories do not cluster well. Moreover, if they are piecewise linear functions, they are mapped to very high dimensional points, further reducing the efficiency of the index. A second direction is to kinetize static indexing structures. The notion of *kinetic data structures*, introduced by Basch *et al.* [5], has led to several interesting results related to moving objects, including results on kinetic space partition trees (also known as cell trees) [2], and on maintaining the Voronoi diagram of a set of moving objects [9]. In the static case, the Voronoi diagram can be processed into a point-location data structure that efficiently answers nearest-neighbor queries. However, no efficient data structure is known for answering point-location queries in a planar sub-division that is continuously deforming. Agarwal *et al.* [1] developed an index that answers a Q2 query in optimal number of I/Os — provided that the queries arrive in chronological order. Since the index maintains as invariant the sorted order of points along one axis, the number of updates in the worst case is $\Omega(n^2)$. To alleviate this problem, one can parametrize a structure such as the R-tree, which partitions the points but allows the bounding boxes associated with the children of a node to overlap. To expect good query efficiency, the areas of overlap and the areas of the bounding boxes must be small. Maintaining these properties over time is likely to be significantly less expensive than maintaining the stronger invariants that other indexes require. Moreover, the R-treeworks correctly even when the overlap areas are too large, although in this case the query performance deteriorates. A kinetic index called TPR-tree, based on the R*-tree, was proposed by Šaltenis *et al.* [11] to handle range queries over moving points. They parametrize the bounding boxes associated with nodes in the R*-treeas follows. Since it is very expensive to maintain the minimum bounding box at all times, they provide a heuristic so that the coordinate of each vertex is a linear function of time. Although initially this heuristic gives a good approximation of the bounding box, it starts to deteriorate with time. To alleviate the problem, whenever the position of a point p is updated, they recompute the boxes on the nodes along the path to the leaf at which p is stored. Although these frequent updates keep the query time low, they have a large overhead.

Our results. We present a *spatio-temporal self-adjusting* R-tree, called STAR for brevity, which is a fully dynamic, R*-tree-based indexing technique. Our approach is similar to TPR-trees, but we introduce the notion of self-adjusting to our index, which allows it to adapt itself without any input from the user, whenever the query performance deteriorates. Our index provides tradeoffs between various performance parameters. A user can specify the parameter that determines the quality of the parametrized bounding box stored at each node of the tree — a better parametrization requires more space. The index also provides a tradeoff between the time spent in self adjusting the structure and the query performance.

3   Indexing Technique

In this section we describe our index STAR-tree. Although it works for points in any dimension, we focus on points in $\mathbb{R}^2$. We first describe the overall structure of the index and the algorithm for computing the parametrized bounding box at each node. We then describe the events at which the index is updated and the procedures for handling the events and for answering queries.

3.1   Overall Structure

A STAR-tree is a B-tree $\mathcal{T}$ whose structure is similar to an R-tree. We will use $\mathcal{T}(t)$ to denote the tree at time t. The points are stored in the leaves. More exactly, if the trajectory of a point p_i is a linear function $p_i(t) = a_i + b_i t$, we store the coefficients $a_i, b_i \in \mathbb{R}^2$ at the leaf. For an internal node v, let $S_v(t)$ be the subset of points stored in the leaves of the subtree rooted at v at time t. For each internal node v, we store a pointer to each of its children w and a compact representation of a parametrized bounding box $\mathcal{B}_w(t)$ that contains all the points in $S_w(t)$ for all $t \geq now$. Let $\text{MBB}_w(t)$ be the minimum bounding box containing $S_w(t)$. For each $t \geq now$, the bounding rectangle $\mathcal{B}_w(t)$ contains $\text{MBB}_w(t)$. The quality of the approximation of $\mathcal{B}_w(t)$ to $\text{MBB}_w(t)$ is controlled by a user-specified approximation factor ε. The better the approximation, the larger the space needed to store $\mathcal{B}_w(t)$. Hence the fanout of the tree becomes smaller, since the total size of a node is fixed (e.g., the disk-block size). Our method allows a trade-off between the fanout of the tree and the approximation factor for the minimum bounding boxes of the nodes. We also maintain certain additional information at each node, to be described below, that determines when some of the nodes need to be reorganized.

We fix the initial time to $t_{start} = 0$. We start by bulk-loading a Hilbert R-tree based on the positions of the moving points at t_{start}. We now describe the procedures for maintaining the bounding boxes at each node and for organizing the tree.

3.2   Computing the Parametrized Bounding Box

For each node $v \in \mathcal{T}$, $\mathcal{B}_v$ is parametrized as a piecewise-linear function. That is, we compute a sequence of intervals $\mathcal{I}_v^x(\tau_1^x), \mathcal{I}_v^x(\tau_2^x), \ldots$ along the x-axis, and a sequence of intervals $\mathcal{I}_v^y(\tau_1^y), \mathcal{I}_v^y(\tau_2^y), \ldots$ along the y-axis so that $\mathcal{I}_v^x(\tau_i^x)$ contains the projection of the points in $S_v(\tau_i^x)$ along the x-axis, and $\mathcal{I}_v^y(\tau_j^y)$ contains the projection of the points in $S_v(\tau_j^y)$ along the y-axis, for all i, j. Let $\langle \tau_1, \tau_2, \ldots \rangle$ be the set $\bigcup \tau_i^x \cup \bigcup \tau_j^y$ in sorted order. Then for any $t \in (\tau_i, \tau_{i+1})$ the interval $\mathcal{I}_v^x(t)$ obtained by linear interpolation from $\mathcal{I}_v^x(\tau_i^x)$ and $\mathcal{I}_v^x(\tau_{i+1}^x)$ contains the projection of $S_v(t)$ along the x-axis; a similar result holds for the linearly interpolated interval along the y-axis. Hence, for any t, the box $\mathcal{B}_v(t) = \mathcal{I}_v^x(t) \times \mathcal{I}_v^y(t)$ contains $\mathrm{MBB}_v(t)$. We discuss below how to compute the sequence $\mathcal{I}_v^x(\tau_1^x), \mathcal{I}_v^x(\tau_2^x), \ldots$ so that for each t the interpolated interval $\mathcal{I}_v^x(t)$ is only slightly larger than the minimum length interval that contains the projection of $S_v(t)$ on the x-axis. The computation of the sequence $\mathcal{I}_v^y(\tau_1^y), \mathcal{I}_v^y(\tau_2^y), \ldots$ is similar. This will ensure that $\mathcal{B}_v(t)$ is tight-fitting at any time t. We assume that the set of points $S_v(t)$ remains unchanged for all t. The cases when $S_v(t)$ changes are handled separately in the insertion/deletion and re-balancing procedures.

The problem of computing $\mathcal{I}_v^x(\tau_1^x), \mathcal{I}_v^x(\tau_2^x), \ldots$ is the same as approximating the *extent* of the points in S_v along the x-axis. For simplicity, we describe the algorithm under the assumption that each point is moving with fixed speed, though it works for more general trajectories.

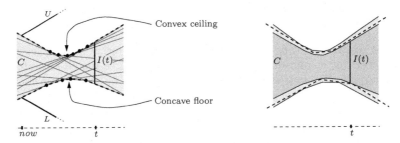

Fig. 2. (i) The interval $I(t)$ is the vertical segment connecting the upper and lower boundaries of C at time t. The rays U, L denote the approximation to $I(t)$ as computed by Jensen *et al.* [11]. (ii) An approximation of C and polygonal chains γ^-, γ^+.

Let $I(t)$ denote the extent of $S_v(t)$ along the x-axis, and let $C = \bigcup_t I(t)$. Then C is a corridor in the tx-plane with convex ceiling and concave floor; see Figure 2. Let ε be the approximation factor. We want to maintain an interval $J(t)$ so that $I(t) \subseteq J(t) \subseteq (1 + \varepsilon)I(t)$ for all t. To do this, we compute two polygonal chains γ^-, γ^+ with the following property: Let C' be the corridor formed by $\bigcup_t (1 + \varepsilon)I(t)$. The chain γ^- lies between the lower boundaries of C and C', and the chain γ^+ lies between the upper boundaries of C and C', respectively. Moreover, γ^- and γ^+ have only $O(1/\sqrt{\varepsilon})$ vertices. The computation of γ^- and γ^+ is done using the greedy algorithm of [3], which guarantees the above properties. The approximate extent is defined by $J(t) = [\gamma^-(t), \gamma^+(t)]$; see Figure 2 (ii). The sequence $\tau_1^x, \tau_2^x, \ldots$ is the set of t-coordinates of the vertices of γ^- and γ^+, and $\mathcal{I}_v^x(\tau_i^x) = J(\tau_i^x)$. By contrast, Jensen et al. [11] approximate $I(t)$ by two rays (Figure 2 (ii)), and the approximation can become very large over time, decreasing the query efficiency of the index. Our experiments indicate that in practice the number of vertices of γ^-, γ^+ is small (between 3 and 12 for $100,000$ points and $\varepsilon = 0.1$). Thus, we can explictly store γ^-, γ^+ at a node of $\mathcal{T}$. If we are unlucky, then γ^-, γ^+ could have too many vertices. In order to ensure that the fanout of a node is not too small, we set a threshold parameter $\lambda \geq 2$ and store only at most λ vertices of each of the four chains. Suppose $t_b(v)$ is the minimum t-value of the last vertex of a chain stored for $\mathcal{B}_v$. Then the parametrization of $\mathcal{B}_v$ is valid only for $t \leq t_b(v)$. We recompute $\mathcal{B}_v(t)$ for $[t_b(v), \infty]$.

If v is an internal node, we do not compute $\mathcal{B}_v$ directly from S_v, but rather in a bottom-up manner from $\mathcal{B}_{w_1}, \ldots, \mathcal{B}_{w_k}$, where $w_1, \ldots, w_k$ are the children of v. There are two issues that one has to handle in this recursive computation. Since $\mathcal{B}_v$ is being computed by approximating the (approximate) bounding boxes computed at it children, the errors accumulate. So one has to choose a smaller value of ε at each node, roughly $\varepsilon/\text{depth}(\mathcal{T})$. Second, the bounding-box representation at the children of v might have been clipped, so $\mathcal{B}_v(t)$ is valid only for $t \leq \min_w t_b(w)$, where w is a child of v. Therefore we set $t_b(v) = \min_w t_b(w)$ if the number of vertices in the chains of $\mathcal{B}_v$ is at most λ. Otherwise, if ξ is the minimum t-coordinate at which a chain of $\mathcal{B}_v$ is clipped then we set $t_b(v) = \min\{\xi, \min_w t_b(w)\}$.

3.3   Events and Event Queue

There are two types of events — external and internal that casue the index $\mathcal{T}$ to be updated. External events are insertion/deletion of a point, and change in the trajectory of a point. Due to lack of space, we do not detail the procedures for handling external events. Roughly speaking, insertions and deletions are handled as in the static R-tree, by looking at the snapshot of the index at the time of insertion or deletion. Changing the trajectory of a point usually requires updating the parametrized bounding boxes along the path to the leaf that stores the point. Besides the external events, there are two types of internal events. Recall that at each node $v \in \mathcal{T}$, $\mathcal{B}_v$ is valid only until $t_b(v)$, so if $t_b(v) < \infty$, we have to recompute $\mathcal{B}_v$ at $now = t_b(v)$. We refer to this event as a box event. As mentioned earlier, the box events are rare even for small values of ε. Our construction maintains the invariant that $t_b(w) \geq t_b(v)$ for any child w of v.

The second internal event is called a conflict event. A conflict event occurs at a node v if the bounding boxes of its children overlap too much. Let $u_1, \ldots u_k$ be the children of v. We replace them with new nodes $w_1, \ldots, w_k$ so that the set of grandchildren of v remains the same and the overlap among the bounding boxes of $w_1, \ldots, w_k$ is smaller than the overlap among those of $u_1, \ldots, u_k$; see

Figure 3. That is, we redistribute the grandchildren of v among its children in order to reduce the overlap among the bounding boxes of the children of v.

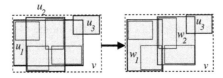

Fig. 3. A conflict event; the children of u_1 nd u_2 are redistributed between w_1, w_2.

There are many ways to define when sibling nodes overlap too much, as well as to compute the new nodes $w_1, \dots, w_k$. We have experimented with a number of heuristics, all of which try to stay close to the R-tree desideratta that the areas of nodes and the areas of overlaps among sibling nodes should be as small as possible. The problem is hard in our case because the area of $\mathcal{B}_v(t)$ changes with time. For each child w_i of v, we compute the relative area of $\mathcal{B}_{w_i}(t)$ that overlaps with the bounding boxes of its siblings. If a point of $\mathcal{B}_{w_i}(t)$ is covered by multiple boxes, it is counted multiple times. We set an area threshold parameter A_v and compute the overlap for a node only at the times when its area becomes $(1+\alpha_v)A_v, (1+2\alpha_v)A_v, \dots, (1+10\alpha_v)A_v$ (α_v is a parameter that we discuss in the experimental section). This approach is related to the intuition that overlaps are only significant for large enough boxes. We set a threshold parameter μ and say that a *conflict* occurs at v at time t if this relative overlap exceeds μ. We set $t_c(v) > now$ to be the earliest time at which a conflict occurs. The value of the area threshold A_v is changed with time because points might move away from each other and thus the areas of bounding boxes increase. We reset A_v every time a conflict event is processed at v. An additional option is to reset this value after each time an external event occurs in a subtree rooted at v. The parameter μ is a ratio between the overlap area and the area of a node. In our experiments, we found that it is best to keep μ unchanged over time.

Finally, we define the *event time* of a node v as $\tau(v) = \min\{t_b(v), t_c(v)\}$. If box and conflict events occur simultaneously, then the box event has higher priority. We maintain a global event queue $\mathcal{Q}$, as a priority queue, that stores the pairs $(\tau(v), e(v))$ for all nodes $v \in \mathcal{T}$ for which $\tau(v) < \infty$. We also use $\mathcal{Q}$ to handle *external events* that are going to occur in the future.

3.4   Answering Queries

As mentioned in Section 2 we focus on three types of queries. Since Q1 and Q2 queries are special cases of Q2' queries, we describe the procedure only for the latter. Let R_1, R_2, t_1, t_2 be the given rectangles and time stamps. Let $\mathbf{R}$ be the volume swept by the query rectangle during the interval $[t_1, t_2]$, i.e., $\mathbf{R} = \bigcup_{t=t_1}^{t_2} R(t)$, where $R(t)$ is defined in Section 2; $\mathbf{R}$ is a convex polytope (see

Figure 1 (c)). We traverse $\mathcal{T}$ in a top-down manner. At each visited leaf z, we report a point $p \in S_z$ if the line segment $\bigcup_{t=t_1}^{t_2} p(t)$ intersects $\mathbf{R}$. At each internal node v, we recursively visit a child w_i of v if $\mathbf{R}$ intersects $\mathbf{B} = \bigcup_{t=t_1}^{t_2} \mathcal{B}_{w_i}(t)$.

Queries Q3 are processed as follows. Let q be the query point for which we want to report the nearest neighbors at time t_q. We describe an algorithm that returns the ε-approximate k-nearest neighbors of q, for any $0 \leq \varepsilon$ and $k \geq 1$. For any node v in the STAR-tree, we define the distance between q and v to be $d(q, \mathcal{B}_v(t))$, if q is outside $\mathcal{B}_v(t)$, and 0 otherwise. The query procedure works in two stages. In the first stage it traverses the index in a top-down fashion and at each level i it maintains a set of nodes $\mathcal{L}_i$ as follows. $\mathcal{L}_0$ is the root . For all $i \geq 1$, $\mathcal{L}_i$ is the subset of children of nodes in $\mathcal{L}_{i-1}$ whose distance to q is minimum over all children of nodes in $\mathcal{L}_{i-1}$. Let h denote the leaf level of the STAR-tree, and $\mathcal{L}_h$ be the subset of leaves corresponding to the recursive definition above. We compute the k nearest neighbors of q among all points stored in the leaves of $\mathcal{L}_h$. Assuming that no two nodes on the same level of the STAR-tree overlap and that the minimum side of each bounding box on that level has length $\Omega(\Delta)$, one can prove by a packing argument that the above procedure visits only a constant number of nodes on each level. To describe the second stage, let Δ_k be the distance from q to the kth nearest neighbor candidate computed during the first phase. Starting from the root, we traverse the tree in a depth-first order, using the following criterion. Let v be the current node, so that v is not a leaf. For each child w of v, if $d(q, \mathcal{B}_w(t)) < \Delta_k/(1 + \varepsilon)$ then visit w recursively. A fast heuristic is to only execute the first stage. This is expected to report points close to k-nearest neighbors but not guaranteed to do so. In our experiments, the heuristic performed very well (see Section 4.5).

4   Experimental Results

The experiments were performed on a Pentium III 800-MhZ machine with 512MB memory, running Linux. The algorithms are implemented on top of the *Transparent Parallel I/O Programming Environment* [4], a templated library that supports efficient high level implementations of external memory algorithms. The entire data and the index structure reside on disk, with the exception of 200K of cached blocks. The cache implements a LRU replacement policy. The block size is 8192 bytes, which, for two dimensional data, leads to a packing of 227 data points per leaf. By contrast, the TPR-tree uses a block of size 4096 bytes, and the packing parameter is 204 (this is because we store coordinates of type double, and the TPR-tree stores coordinates of type float). Thus, even though we use larger blocks, the leaf fanout and the number of leaves are similar in both structures. Because both the STAR-tree and the TPR-tree have very small height, the upper levels of both structures contain very few nodes, and thus the larger block size in our experiments does not significantly influence the results. Taking into account all these reasons, the performance values obtained with the STAR-tree on the same data and experimental set-up used by the TPR-tree are a good indication on the relative performance of the two indexes.

Parameter choices. In all the experiments $\lambda = 4$, which leads to a fanout of 30 for internal nodes. We set $\varepsilon = 0.1$. The value α_v depends on the level of the node v. The higher the node is (with the root being highest), the larger α_v is. Hence, we check for overlap less frequently at higher levels in the tree, and thus we do not reorganize upper nodes too often. We have determined experimentally that choosing $\alpha_v = 0.05 \cdot level(v)$ results in a reasonably small number of overlap events, while maintaining good query performance over time. The area A_v is reset after a conflict among children of v is solved, as well as after a child of v is deleted. We do not reset A_v after a child is inserted in v. Each time, we set A_v to be the largest area of a child of v at the time of the reset. The overlap threshold μ is set during the initialization of the node and it does not change over time. In Section 4.1 we evaluate the performance of the STAR-tree for range queries (the only ones for which the TPR-tree performance is reported). We then present experimental results for the nearest neighbor queries in Section 4.5.

4.1   Range Query Results

For all experiments reported in this subsection, we perform a combination of queries of types Q1, Q2, and Q2'. The relative proportions of these queries are 0.6, 0.2, and 0.2, the same as in [11]. We use three types of data. The first type was generated using the generator provided by the authors of the TPR-tree, and we use it for comparison purposes. The second type of data sets are similar to the first set in the distribution of initial positions and trajectories, but generate significantly fewer updates, and allow points to appear and disappear from the data. Finally, we provide experimental results on a realistic data set generated by using information on NC roads extracted from the TIGER/Line data of the US Bureau of the Census.

4.2   Comparison with the TPR-Tree

We generated $100,000$ points using the data generator provided by the authors of the TPR-tree (see [11] for details). The number of destinations is ND = 20. In addition to the initial positions of the points, the generator also outputs a very large number of updates that must be performed by the structure during the course of the simulation. For $100,000$ points, there are close to 1 million point updates, each consisting of two operations: a deletion followed by a re-insertion. Although the updates occur with varying frequencies during the simulation, the average frequency is 7 updates per node per time unit. Hence, the updates act, in effect, as a time-sampling mechanism that allows the indexes to adjust their information very frequently. We ran two experiments: one in which we store the exact bounding box for each node, and one in which we store the approximate box. We report the performance of the STAR-tree in Figure 4 (a). This should be compared to the results reported in [11], Figure 15, on a data set generated with the same parameters. Since we do not have the exact numbers, we chose not plot the performance of the TPR-tree on the same graph. However, we note that, as reported by the authors, the TPR-tree requires an average of about 65 I/O's per query after time $t \geq 360$. Hence, we achieve a speed-up of 3 using exact boxes, and of 2 using approximate boxes. In the case of exact boxes, the overall number of internal events is less than 1.4% of the number of external events (i.e. trajectory updates). When we maintain approximate boxes, there is only one update event and a similar number of overlap events. We conclude that, with only a small overhead, the STAR-tree is able to adjust itself and maintain better performance over time.

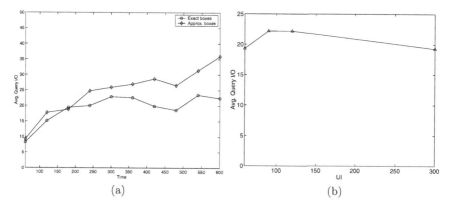

(a)   (b)

Fig. 4. Performance of STAR-tree under frequent updates: (a) over time; (b) for different UI.

The implementation of the STAR-tree does not require having an estimate on the time horizon for which the tree is in use. This is important for applications on which it is hard to estimate the horizon. For the TPR-tree, the time horizon H is proportional with the update interval UI, i.e. the average time between two succesive updates of the same point. The value H is then used in the bulkloading and updating of the tree. Hence, changing H affects the performance. Figures 9 and 10 in [11] show that for $UI = 60$ the average query I/O is between 40 and 50, depending on the setting of H. For $UI = 120$, the query performance varies between 60 and 90. In contrast, the STAR-tree has the same performance for different settings of UI; see Figure 4 (b).

4.3   Performance of STAR-Tree under Few Updates

In this subsection we study the performance of the STAR-tree when the frequency of updates is significantly lower than for the datasets of the previous section. For example, in applications such as airline traffic, points tend to maintain their velocities and directions for a long period of time. We generate new data sets that closely resemble the previous sets in terms of the distribution of points at loading time, as well as the distribution of trajectories. The significant difference is that no updates are generated while the point travels on a leg between two destinations. We also allow insertions and deletions of the points. We insert 10% of the data points after the initialization, and we delete 10% of the points before the simulation ends. The insertion and deletion times are randomly distributed throughout the simulation. The overall number of external events that the index receives is 11 times smaller than for the previous data set. The data consists of 100, 000 points. We perform experiments with the number of hubs ND set to 20 (very skewed data), and to 1, 000 (skewed data). We denote by S_1 the first set, and by S_2 the second set. We also denote by S_3 the set of uniform data.

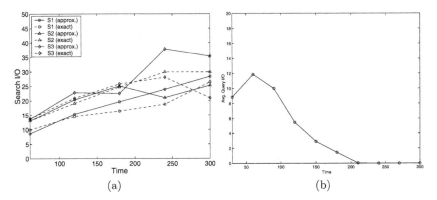

Fig. 5. Performance of STAR-tree under few updates: (a) skewed and uniform data; (b) realistic Durham data.

Figure 5 (a) shows that the performance of STAR-tree does not deteriorate too much over time. The uniform dataset is the only one for which a significant performance loss appears between $t = 180$ and $t = 240$. However, the index readjusts itself afterwards and the performance improves for $t \geq 240$. For each of S_1, S_2, and S_3, two sets of results are plotted. The first set is for the case when we store the exact bounding boxes, and the second for approximate boxes with $\varepsilon = 0.1$. In the second case, approximate bounding boxes are generated only if storing the chains of the exact boxes requires more than λ vertices per chain. Note that the usage of approximate bounding boxes only marginally worsens the performance for sets S_1 and S_2, and affects the set S_3 only after a significant amount of time. However, the number of box events drops dramatically (e.g., for the set S_1, it drops from 2667 to 9).

The effect of varying the query window size is shown in Figure 6 (i). The average is taken over a simulation time of 300 for set S_1. Finally, Figure 6 (ii) shows the influence of data size on the average query performance. All sets we use are uniform, because this is the case in which the index has worst performance. The averages are for a simulation time of 300.

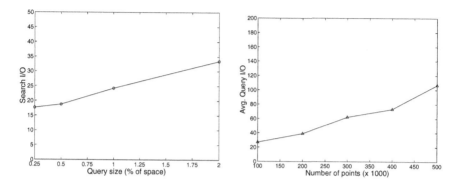

Fig. 6. (i) Effect of query size on the search performance of the STAR-tree. (ii) Search performance of the STAR-tree for varying number of points.

4.4   Realistic Traffic Data

We have generated $51,000$ data points that model car traffic as follows: We extracted the roads map around Durham, NC (within a square of 120 miles centered at Durham) from the National Transportation Atlas Data provided by Bureau of Transportation Statistics (see `http://www.bts.gov/gis/ntatlas/natlas.html`). This resulted in a a set of $231,554$ polygonal chains describing the roads around Durham. Using this information we computed a planar-map representing the road map of the region. To generate a realistic path, we randomly selected two points on the map, and computed (using Dijkstra's algorithm) the shortest path on the map between those points. We then simplified each path using the Douglas-Peucker heuristic [7] so that the simplified path has at most five vertices (to avoid frequent trajectory updates). The simplified path is typically almost identical to the original path. The routes are generated in three length ranges (relative to the diameter of the graph). A car is then generated to travel along the route, and its speed varies according to the length of the route. When a car reaches the final point on its route, it signals the index that it should be deleted. Figure 5 (b) shows how the performance of the search changes over time. The experiment was run using approximate bounding boxes with $\varepsilon = 0.1$. Because of the cache, the average I/O for the last 60 time units is 0.

4.5   Nearest Neighbor Query Results

During each time unit, we generate four nearest neighbor queries uniformly at random over the entire data space. Unless otherwise specified, each simulation lasts 200 time units. The data sets are the same from the previous section.

Accuracy results. We analyze the accuracy with which the heuristic method answers nearest neighbor queries using various measures. In Figure 7 (a) we show results for 1-nearest neighbor queries. The solid bars show the percentage of queries for which our procedure returns the exact 1-nearest neighbor. The hashed bars represent the percentage of queries for which the nearest neighbor as computed by our procedure is a 0.1-approximate nearest neighbor. Finally, the grid-patterned bars represent the percentage of queries for which the nearest neighbor computed by our procedure is one of the exact 10-nearest neighbors. For sets $S2$ and $S3$, these percentages are in the range of $98 - 99\%$, while for $S1$ and *Durham* they are lower, ranging from 80% to 96%.

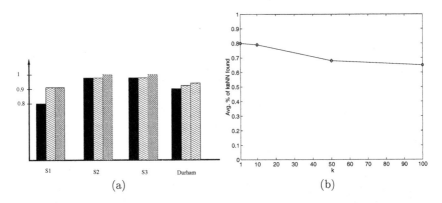

Fig. 7. Accuracy of heuristic method for: (a) 1-nearest neighbor; (b) k nearest-neighbor.

For each data set, we also compute the accuracy of answering k-nearest neighbor queries, for $k = 10$. For each query, we determine how many of the 10 points we return are among the exact 10-nearest neighbors of the query, and we average this number over all queries. Our experiments show that for $S2$ and $S3$ this average is about 9.8, implying that for almost all queries we return the exact 10-nearest neighbors. For $S1$, the average is 7.9, and for *Durham* it is 8.7. However, the accuracy in terms of the average percentage of exact k nearest neighbors found degrades as k increases. Figure 7 (b) illustrates this behavior for set $S1$. We report the average (over all queries) percentages of exact and 0.1-approximate nearest neighbors that the heuristic finds.

Efficiency results. We study the efficiency of the exact, approximate and heuristic methods in terms of the average number of I/O's required to answer a query. The heuristic method is always faster, since we only execute one traversal of the index (see the previous section). The exact method is at most as fast as the approximate method. We report our results both in the presence and in the absence of the cache. As we show below, the presence of the cache causes all three methods to have similar efficiency. This is to be expected, since some nodes visited during the second traversal are likely to have been cached during the first traversal.

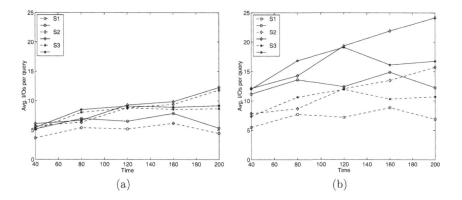

Fig. 8. I/O efficiency for (a) cache present; (b) cache not present.

We have run experiments for $k = 1$, 5, and 10, and for $\varepsilon = 0$, 0.1, 0.2, and 0.4, and have concluded that the average number of I/O's does not change significantly for these values of k and ε. This indicates that the 10 nearest neighbors of most queries are stored together. Moreover, in the absence of the cache, our experiments show that the average number of I/O's for a query is not much larger than the depth of the tree (for the heuristic method), or twice the depth of the tree (for the other two methods). This is an indirect indication that the STAR-tree has small overlap among nodes on the same level. Figure 8 shows experimental results for $k = 10$ for sets S_1, S_2 and S_3 both in the presence and in the absence of the cache, for the heuristic (dashed lines) and exact methods (solid lines). Results for the approximate method (with $\varepsilon = 0.1$) are close to those for the exact method. We report the average query I/O every 40 time units. We expect that the efficiency of the exact and approximate methods will decline when k becomes too large. The number of I/O's in the heuristic method does not depend on k. Hence, its efficiency will remain the same. However, its accuracy in terms of the number of exact k nearest neighbors found will degrade. Due to lack of space, we do not report experiments for the realistic data set (since this set is smaller, it should not be reported together with the other sets). However, these experiments show that the index adapts itself as points are deleted, and the query performance improves after 100 time units.

5 Conclusions

We presented a new technique called STAR-tree for indexing moving points in the plane. Our experiments clearly demonstrate the advantages of our technique. By intergrating a few geometric techniques with the indexing techniques, we were able to circumvent the problem of updating the index frequently and to improve the query performance. The next step would be to handle uncertainty, to answer more complex queries, and to incorporate a few learning techniques in the self-adjusting procedure.

Acknowledgments. The authors thank Jeff Erickson, Christian Jensen, and Ouri Wolfson for various useful discussions, Simonas Šaltenis for answering several questions related to his paper [11], Simonas and Christian for providing their data generator, and Jan Vahrenhold for providing his R-tree code.

References

1. P. K. Agarwal, L. Arge, and J. Erickson, Indexing moving points, *Proc. Annu. ACM Sympos. Principles Database Syst.*, 2000. 175–186.
2. P. K. Agarwal, J. Erickson, and L. J. Guibas, Kinetic BSPs for intersecting segments and disjoint triangles, *Proc. 9th ACM-SIAM Sympos. Discrete Algorithms*, 1998, pp. 107–116.
3. P. K. Agarwal and S. Har-Peled, Maintaining approximate extent measures of moving points, *Proc. 12th ACM-SIAM Sympos. Discrete Algorithms*, 2001.
4. L. Arge, R. Barve, D. Hutchinson, O. Procopiuc, L. Toma, D. E. Vengroff, and R. Wickeremesinghe, *TPIE User Manual and Reference (edition 0.9.01b)*. Duke University, 1999. The manual and software distribution are available on the web at http://www.cs.duke.edu/TPIE/.
5. J. Basch, L. J. Guibas, and J. Hershberger, Data structures for mobile data, *Proc. 8th ACM-SIAM Sympos. Discrete Algorithms*, 1997, pp. 747–756.
6. N. Beckmann, H.-P. Kriegel, R. Schneider, and B. Seeger, The R*-tree: An efficient and robust access method for points and rectangles, *Proc. ACM SIGMOD Conf. on Management of Data*, 1990, pp. 322–331.
7. D. H. Douglas and T. K. Peucker, Algorithms for the reduction of the number of points required to represent a digitized line or its caricature, *Canadian Cartographer*, 10 (1973), pp. 112–122.
8. R. H. Güting, M. H. Böhlen, M. Erwig, C. S. Jensen, N. A. Lorentzos, M. Schneider, and M. Vazirgiannis, A foundation for representing and querying moving objects, *ACM Trans. Database Systems*, 25 (2000), pp. 1–42.
9. M. I. Karavelas and L. J. Guibas, Static and kinetic geometric spanners with applications, *Proc. 12th Annu. ACM-SIAM Sympos. Discrete Algorithms*, 2001, pp. 168–176.
10. G. Kollios, D. Gunopulos, and V. J. Tsotras, On indexing mobile objects, *Proc. Annu. ACM Sympos. Principles Database Syst.*, 1999, pp. 261–272.
11. S. Šaltenis, C. S. Jensen, S. T. Leutenegger, and M. A. Lopez, Indexing the positions of continuously moving objects, *Proc. ACM SIGMOD International Conference on Management of Data*, 2000, pp. 331–342.

12. A. P. Sistla and O. Wolfson, Temporal conditions and integrity constraints in active database systems, *Proceedings of the 1995 ACM SIGMOD International Conference on Management of Data*, 1995, pp. 269–280.

13. A. P. Sistla, O. Wolfson, S. Chamberlain, and S. Dao, Modeling and querying moving objects, *Proc. Intl Conf. Data Engineering*, 1997, pp. 422–432.

14. J. Tayeb, O. Ulusoy, and O. Wolfson, A quadtree-based dynamic attribute indexing method, *The Computer Journal*, (1998), 185–200.

An Improvement on Tree Selection Sort[*]

Jingchao Chen

Bell Labs Research China, Lucent Technologies
3/F, Aero Space Great Wall Building No.30,
Hai Dian Nan Lu, Beijing, 100080, P.R.China
chenjingchao@yahoo.com

Abstract. The standard Tree Selection Sort is an efficient sorting algorithm but requires extra storage for n-1 pointers and n items. The goal of this paper is to not only reduce the extra storage of Tree Selection Sort to n bits, but also keep the number of comparisons at $n\log n + O(n)$. The improved algorithm makes at most $3n$ data movements. The empirical results show that the improved algorithm is efficient. In some cases, say moving one item requires at least 3 assignment operations, the algorithm is the fastest on average among known fast algorithms.

1   Introduction

Tree Selection Sort   (also known as Tournament Sort, because it resembles the principle of tournament matches) was discovered in the fifties [4,5], and it has an optimal number of comparisons of $n\log n + O(n)$. However, because it needs extra space for n-1 pointers and n output items (when n is a power of 2), the research on it was discontinued by its successor algorithm which was christened Heapsort by its inventor Williams [1]. Heapsort takes $2n\log n$ comparisons, but uses no extra space. In order to optimize the number of comparisons made by Heapsort, one has made a substantial effort on improving Heapsort, including the work of Floyd [2], Gonnet and Munro [9], Carlsson [8], McDiarmid and Reed [7], Wegener [6] and Dutton [10]. Particularly, Weak-heap Sort [10] by Dutton improves the time complexity of Heapsort, and attains a worst case number of comparisons of $n\log n + 0.086n$. However, it uses n extra bits, and requires $O(n\log n)$ movements. By the empirical results of [11], Weak-heap Sort used about $0.5n\log n$ movements on average.

   Both Proportion Split Sort [11] and Proportion Extend Sort [12] are more efficient than the best-of-three version of Quicksort [3], and make $O(n\log n)$ comparisons in the worst case. The empirical results reveal that they used about $0.25n\log n$ movements on average when the split factor p of Proportion Split Sort is 24 [11] and the extension factor p of Proportion Extend Sort is 16[12].

---

* This work was partially done while the author served in Shanghai Jiaotong University, China, and partially supported by the National Natural Science Foundation of China grant 69873033.

D. Mount and C. Stein (Eds.): ALENEX 2002, LNCS 2409, pp. 194–205, 2002.
© Springer-Verlag Berlin Heidelberg 2002

Theoretically, any sorting algorithm requires at least log n! (=$n\log n$-1.44n) comparisons and n movements. If we want to implement such an algorithm approaching these two lower bounds, How much extra space does it require? Mergesort is one of the algorithms challenging these two lower bounds, but if the number of movements is limited to be n, then it needs at least $n\log n$ extra bits. As described previously, although the standard version of Tree Selection Sort is able to approach these two lower bounds, it needs more extra space than Mergesort.

To improve on Tree Selection Sort, we propose four strategies. The first strategy, given in Section 2, is to eliminate the output area (n items of space) required by a naive implementation. The second strategy, given in Section 3,can reduce the extra space to $2n$ bits by storing offsets rather than complete pointer. The third strategy, given in Section 4, can bring the extra space down to $2n/m$ bits. Because the usual computer does not support bit access operations, the algorithm described by the third strategy is not very efficient. For this reason, the fourth strategy, given in Section 5, expands storage space for a part of the nodes and compress further storage space for another part of the nodes. Section 6 describes the simulations to measure the performance of our improved algorithm and the other fast algorithms.

2   Tree Selection Sort without an Output Area

This section is divided into four subsections. The first subsection will present a basic improvement on Tree Section Sort. The other subsections will analyze its time and space requirement.

2.1   A Basic Improvement on Tree Section Sort

Similar to the standard Tree Selection Sort, The basic version of our algorithm employs also a tree called a global-pointer tree but removes the output area of the standard version. Next we define formally a global-pointer tree as a binary tree where:

(1) the value in the i-th leaf is the i-th item of the sequence to be sorted, and
(2) the value in any internal node is a pointer or index specifying the position of the largest value of all the leaves in its subtree.
    Figure 1 shows a global-pointer tree with 8 items.
    In the creation phase of a pointer tree, both our algorithm and the standard version are same. The difference is that our algorithm always puts the current maximum into the correct position in input area in the selection phase. For convenience, here we call the basic version of our algorithm global-pointer-tree-sort. This algorithm may be described as follow.
    global-pointer-tree-sort (A, n)
    Suppose that the items to be sorted are stored in A[0], A[1], ... , A[n-1].
(1) Creation phase of a global-pointer tree
According to the principle of tournament matches, we may create bottom-up a global-pointer tree as shown in Fig.1 by the following code:

for $i=1, \ldots, H$:
    for $j=0, \ldots, N_i$:
        if $A[P_{(i-1)2j}] < A[P_{(i-1)(2j+1)}]$ **then** $P_{ij} = P_{(i-1)(2j+1)}$
        else $P_{ij} = P_{(i-1)2j}$

where H is the height of the tree, $2N_i + 1$ is the maximal number of the nodes at height i, P_{ij} is the value in the j-th internal node at height i, and $P_{0j} = j$ for $0 \le j < n$.

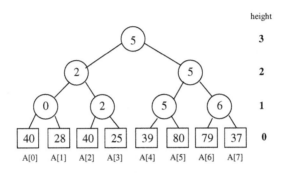

Fig. 1. A global-pointer tree

(2) Selection phase
We use the leaves to store the sorted output sequence in a manner similar to Heapsort. In more detail suppose $A[0..m]$ holds the input we are currently working on and $A[max]$ is the maximum of these elements, so we want to output it into $A[m]$. If $A[m]$ is the smaller element in its first comparison, simply swap $A[max]$ and $A[m]$, and then repair the tournament along $A[m]$'s new path. Otherwise if $A[m]$ is the larger element in its first comparison, swap it with the element it was compared to, repair the tournament above the comparison node, and then proceed as in the first case. The procedure may be described as follows.

    for $m=n-1, \ldots, 1$:
        if $m=max$ **then** `UpdateLargestPath(`m`-1)`
        else `UpdateRightmostPath(`m`)`
            `Exchange A[`m`] and A[`max`]`
            `UpdateLargestPath(max)`

Figure 2 shows a scenario of the procedure. Routines UpdateRightmostPath and UpdateLargestPath given above update each node in the path from leaf m to the root and the path from leaf max to the root, respectively, because the values in the related nodes are incorrect after leaf max is exchanged with leaf m.

UpdateRightmostPath guarantees that after removing the m-th leaf, the tree formed by the remaining $m-1$ leaves is still a global-pointer tree defined earlier. Suppose the set of the nodes on the path from the m-th leaf to the root is $\{R_1, R_2, \ldots, R_t\}$ where R_1, R_{k+1} ($0 < k < t$) and R_t are the m-th leaf, the parent of R_k and the root, respectively, as shown in Fig.2. The implementation details of UpdateRightmostPath may be described as follows:

Procedure UpdateRightmostPath (*m*)

(1) Search bottom-up for the first right child R_k (see Figure 2(a)).

i.e. **for** *k*=1 **to** *t*-1 : **if** R_k is the right child of R_{k+1}, **then** goto (2).

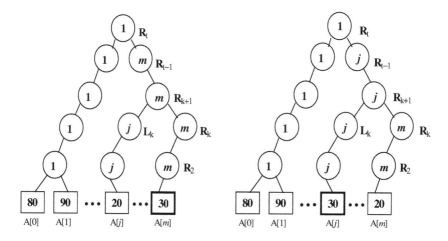

(a) Before running UpdateRightmostPath

(b) After running UpdateRightmostPath

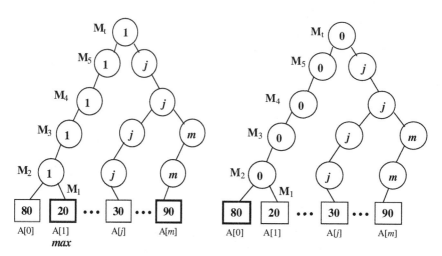

(c) After exchanging A[1] and A[*m*]

(d) After running UpdateLargestPath

Fig. 2. Handling the largest item of A[0..*m*]

(2) **if** $R_{k+1} \neq m$, **then return.**

Let the left sibling of R_k be L_k.

exchange $A[L_k]$ and $A[m]$.

(3) Update the value of the ancestor of R_k.

i.e. **for** $i=k+1, \ldots, t$:

if $R_i \neq m$, **then return.**

$R_i := L_k$

endfor

As can be seen from Fig.2 (c), the largest item of $A[0..m]$ is moved to the correct position if we exchange $A[m]$ and $A[max]$ (note *max* is stored in the root) after running UpdateRightmostPath(m), and the selection of the next-largest item follows. We perform the selection of the next-largest item by procedure UpdateLargestPath. Suppose the set of the nodes on the path from leaf *max* to the root is $\{M_1, M_2, \ldots, M_t\}$ where M_1, M_{i+1} ($0 < i < t$) and M_t are leaf *max*, the parent of M_i and the root, respectively, as shown in Fig. 2(c) (d). The implementation details of UpdateLargestPath may be described as follows:

Procedure UpdateLargestPath(M_1)

Let the sibling of M_i be B_i

for $i=1$ **to** $t-1$:

if $A[B_i] < A[M_i]$, **then** $M_{i+1} := M_i$

else $M_{i+1} := B_i$

Now it is easy to see that the entire array A is sorted in ascending order.

2.2 Number of Comparisons Required in the Worst Case

In the creation phase of the tree, we create one internal node using at most one comparison. Thus, we can imply that the number of comparisons to construct a pointer tree is at most n in the worst case. Since no data items are compared in UpdateRightmostPath, and the number of comparisons required by each call to UpdateLargestPath is at most $\log n$, the height of the tree, the selection phase of the algorithm needs at most $n \log n$ comparisons. All together, the number of comparisons required by the algorithm is at most $n \log n + n$.

2.3 Number of Data Movements Required in the Worst Case

No movements occur in the creation phase of the tree, both UpdateRightmostPath and Exchange require at most one movement, and the number of calls of these two procedures is at most n. Hence, the overall number of movements is at most $2n$.

2.4 Space Requirements

The number of internal nodes of a pointer tree is given by $\lceil n/2 \rceil + \lceil n/4 \rceil +, \ldots ,+2 + 1$, where n is the size of the array to be sorted. This sum is about n. Since $\lceil \log n \rceil$ bits are enough to store the value of any internal node, the extra space required by the algorithm is approximately $n \log n$ bits.

In the next two sections, we will improve this basic version further and restrict our attention to space requirements of a pointer tree.

3   Reducing Extra Space by Way of a Local-Pointer Tree

In this section, we reduce the extra space of the sorting algorithm described in the previous section, using the notion of a local-pointer tree defined below.

A binary tree is said to be a local-pointer tree if
(1) the value in the i-th leaf is the i-th item of the sequence to be sorted, and
(2) the value in any internal node is an offset, i.e. the index of the largest leaf of its subtree minus the index of the first leaf of its subtree.

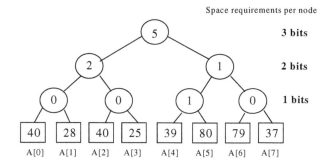

Fig. 3. A local-pointer tree

Figure 3 shows a local-pointer tree corresponding to the global-pointer tree shown in Fig.1. The sorting algorithm based on a local-pointer tree structure   (Local-algorithm, for short) can be obtained if we make a slight modification of the sorting algorithm based on a global-pointer tree structure (Global-algorithm, for short). The difference between the two algorithms is that the Local-algorithm accesses indirectly the item of the array by the value of the internal node, and the Global-algorithm accesses directly the item of the array by the value of the internal node. Suppose L_{ij} denotes the value of the j-th internal node of the local-pointer tree at height i (the height of a node is the number of edges in the path from that node to a leaf on the bottom level of the tree). And let G_{ij} be the index of the largest leaf of the subtree rooted at node L_{ij}. In constructing a local-pointer tree, the Local-algorithm updates L_{ij} by

$$L_{ij} := G_{ij} - j \times 2^i \qquad \text{for } i > 0.$$

In the selection phase, the Local-algorithm obtains G_{ij} by

$$G_{ij} := L_{ij} + j \times 2^i \qquad \text{for } i > 0.$$

Clearly, storing only L_{ij} (not G_{ij}) can implement sufficiently the Local-algorithm. The right-hand side of Fig.3 shows the extra bits required by per node at each level.

Since a node at height i has at most 2^i leaves, i bits are quite enough to store the value of a node at height i. Thus, we may compute the number of extra bits required by the Local-algorithm by

$$1 \times \lceil n/2 \rceil + 2 \times \lceil n/2^2 \rceil + 3 \times \lceil n/2^3 \rceil +, \dots, + \lceil \log n \rceil \times 1.$$

This sum is $2n - 2 - \log n$ when n is a power of 2.

4   Reducing Extra Space by Way of a Multi-fruit-per-Leaf Tree

In order to define vividly the notion of the tree given below, here we view items to be sorted as fruits. A m-fruit-per-leaf tree is formally defined as a local-pointer tree where:

(1) each leaf contains $m(m \geq 1)$ fruits except the last leaf contains m or less than m fruits, and

(2) the value in leaf i means the maximum of its fruits(items) $A[i \times m]$, $A[i \times m+1]$, ... , $A[i \times m+m-1]$.

Figure 4 shows a 2-fruit-per-leaf tree with 4 leaves. This tree corresponds to the local-pointer tree shown in Fig.3. Because the number of leaves in a m-fruit-per-leaf tree is n/m, from the previous computation, we conclude easily that the extra space required by the sorting algorithm based on a m-fruit-per-leaf tree (Fruit(m)-algorithm, for short) is about $2n/m$ bits. When $m=2$, this formula is n. Note that the Fruit (1)-algorithm is exactly the Local-algorithm described above.

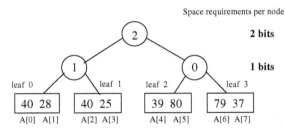

Fig. 4. A 2- fruit-per-leaf tree

The key problem the Fruit (m)-algorithm needs solve is how to compute the value in a leaf. Depending on the method used in this part, we can get various algorithms with different complexities. When m is small, for example, $m < 10$, binary insertion sort is a better choice. When $m=2$, the case becomes simpler.

Next we investigate only the case $m=2$. During the creation phase of a 2-fruit-per-leaf tree, we need to add a preprocessing operation, which puts a larger fruit in the left, and a smaller fruit in the right. This operation needs also to be performed before executing `UpdateLargestPath`. Obviously, these two parts of the Fruit (2)-algorithm adds altogether $1.5n$ data movements in the worst case. An advantage of the Fruit (2)-algorithm is that when the last position is the right, it is unnecessary to execute `UpdateRightmostPath` during the selection phase. This is because the fruit on the right is always smaller one. One call to `UpdateRightmostPath` generates one data movement. Since altogether $n/2$ smaller fruits are stored in the right, this part of the Fruit (2)-algorithm can save $n/2$ data movements. From these analyses, it is easy to see that the overall number of movements required by the Fruit (2)-algorithm is at most $3n$, and the overall number of comparisons required by the Fruit (2)-algorithm is the same as that required by the Local-algorithm, i.e. at most $n\log n+n$.

5 Multi-tier Trees and Non-redundant Trees

To save the storage space, the Fruit (2)-algorithm has to store compactly all internal nodes of a tree in consecutive memory locations in bits. However, the usual computer does not support bit access operations. Thus, there is the possibility the Fruit (2)-algorithm cannot work efficiently on the usual computer. In fact, by empirical testing, we found that the Fruit (2)-algorithm is not very efficient.   For this reason, we introduce the notion of multi-tier tree to relax the limit to the extra space of some nodes by employing the Fruit (32)-algorithm instead of the Fruit (2)-algorithm. In the Fruit (32)-algorithm, it is inefficient to compute the value in a leaf having 32 fruits by Binary Insertion Sort. However, if we compute the value in a leaf by using the Fruit (2)-algorithm, this computation part of the Fruit (32)-algorithm becomes efficient. For convenience, we call the algorithm based on this idea, Fruit [32-2]-algorithm. the Fruit [32-2]-algorithm uses a 32-fruit-per-leaf tree, each leaf of which is a 2-fruit-per-leaf tree. That is, the tree used by this algorithm is a two-tier tree shown in Fig.5. In the 32-fruit-per-leaf tree, all internal nodes are stored in words (16-bits) to speed up this algorithm. Thus, the number of extra bits required by the internal nodes in the 32-fruit-per-leaf tree is $\lceil 1/16 \rceil \times 16 \times \lceil n/64 \rceil + \lceil 2/16 \rceil \times 16 \times \lceil n/128 \rceil +, \dots , + \lceil \log n /16 \rceil \times 16 \times 1$. This sum is about $0.5n$. In each 2-fruit-per-leaf tree, all internal nodes are stored in bits to save the extra space. Thus, the extra space required by a 2-fruit-per-leaf tree with 16 leaves is $8 \times 1+4 \times 2+2 \times 3+1 \times 4=26$ bits. Next we use the notion of non-redundant tree to improve further its space requirement.

A tree is said to be non-redundant if its redundant nodes are removed. If an internal node has the same information as its parent, we call it a redundant node. Figure 6 shows the redundant nodes of the tree shown in Fig.1. Reducing redundant nodes with X's converts Fig.6 (a) into Fig.6 (b). A redundant tree can be converted into a non-redundant tree by way of bottom-up. Conversely, a redundant tree can be reconstructed also from a non-redundant tree by way of top-down. If building a 32-fruit-per-leaf tree with $n/32$ leaves by this trick, half of its extra space can be saved, but such an algorithm becomes complicated. If building a leaf with 32 fruits by this trick, the algorithm supporting it is still simple, and the extra space is reduced to 4×1 + 2×2 + 1×3 + 1×4 = 15 bits.

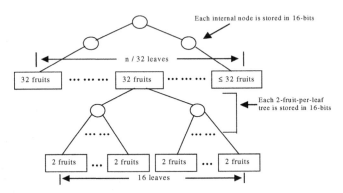

Fig. 5. A two-tier tree

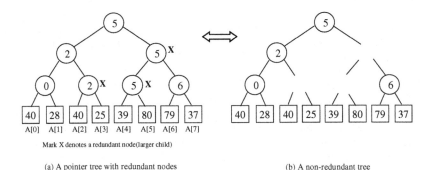

(a) A pointer tree with redundant nodes   (b) A non-redundant tree

Fig. 6. A redundant tree and a non-redundant tree

Based on this analysis, the Fruit [32-2]-algorithm builds each of non-redundant 2-fruit-per-leaf trees, using 16 extra bits. Thus, the extra space required by all leaves in a 32-fruit-per-leaf tree is $n/32 \times 16 = 0.5n$ bits. Therefore, we can conclude that if using the two-tier tree, the overall extra space required by the Fruit [32-2] algorithm is about n bits. It is easy to see that the time complexity of the Fruit [32-2] is the same as that of the Fruit (2)-algorithm given in the previous section.

6 Conclusions and Empirical Results

We made experiments to evaluate the average execution time of our modified algorithm and the other fast algorithms, using three different personal computers:

Pentium II/350MHz, 80386DX/40MHz and Celeron/700MHz. All the algorithms are written in C and run under the MS-DOS operating system.

Tables 1–3 show the average execution time required by the algorithms running on different computers. In these three tables, notations Psort, Esort and Wsort are short for Proportion Split Sort (split factor $p=24$)[11], Proportion Extend Sort (extension factor $p=16$)[12] and Weak-heap Sort [10], respectively. Both Qsort and QsortR are the best-of-three version of Quicksort (also called Clever Quicksort). Qsort chooses always a key as the *pivot* from such three keys: leftmost, middle and rightmost, while QsortR chooses randomly a key as the *pivot* from the set. In [14], Sedgewick suggests that sorting small subarrays should use a faster sorting method, say insertion sort. In our simulations, Psort, Esort, Qsort and QsortR do not adopt Sedgewick's strategy. Fruit[32-2] is the Fruit[32-2]-algorithm given in the previous section. Here we always build the current last leaf in a 32-fruit-per-leaf tree by a redundant 2-fruit-per-leaf tree to speed up the Fruit[32-2]-algorithm, but the other leaves are built by a non-redundant 2-fruit-per-leaf tree.

The first column in Tables 1,2 lists the size per item to be sorted. The last two rows in Table 1, MOVE and COMPARE, reflect the average number of exchanges and comparisons for each algorithm, respectively. Note that all the numbers of bytes per item are even. This is because we assume that the unit of data movements is 2 bytes. That is, if the size of items A, B is $2 \times m$ bytes, the assignment statement A:=B is done by m assignment operations of word-based computers. Each average execution time in Tables 1,2 is based on 60 input sets, each of which consists of 50000 randomly generated distinct integers.

As can be seen in Tables 1, 2, in all cases, Fruit[32-2] is much faster than Wsort. When the size per item is lager than 6 bytes, Fruit[32-2] is the fastest among all the algorithms observed. There is such an interesting phenomenon: when the size per item is 6 bytes, Table 1 shows that Fruit[32-2] is faster than Esort while Table 2 shows that Fruit[32-2] is slower than Esort. The reason for leading to this phenomenon may be that the two computers have different architectures.

In both theory and experiment, A. LaMarca and R.E. Lader have investigated the effect that caches have on the performance of four sorting algorithms: mergesort, quicksort, heapsort, and radix sort [13]. For the limitation of space, this presentation gives only a few simple empirical results on the influence of caches on the performance of our sorting algorithm, rather than analyzing it in details. Table 3 shows the average execution time to sort n 4-byte integers using Celeron/700MHz, which is a computer with caches. As before, for each instance, we generated 60 input sets, each of which consists of n randomly generated distinct integers. Both Qsort and Esort are in-place sorting algorithms. When key lengths are sufficiently large, data movements spend much time. To save the time of movements, the two algorithms use usually index tricks. For this reason, Columns 4 and 6 give the empirical results for the index versions of Esort and Qsort. When keys are required to exchange, the index versions exchange their indexes instead of exchanging the keys. As can be seen in Tables 3, Fruit[32-2] is faster than Wsort, but slower than Esort and Qsort. When n is not so large, say, $n \leq 100000$, Fruit[32-2] is faster than the index version of Esort and Qsort. However, when n is very large, Fruit[32-2] becomes inefficient. This may be due to the influence of caches on its performance.

In the previous section, we have theoretically shown that the improved version of Tree Selection Sort, Fruit[32-2], makes $n\log n+n$ comparisons and $3n$ data movements in the worst case. In this section, by the empirical results, we have revealed that the additional operations supporting this algorithm are not time-consuming. In some cases, this algorithm can compete with the other fast algorithms such as Proportion Extend Sort and Quicksort. Since the influence of caches on the performance of our algorithm is large, our further research will be focused on how to improve the cache locality of our algorithm.

Table 1. Average execution time to sort 50000 integers (in milliseconds, Pentium II/350MHz)

Size per item	Fruit [32-2]	Psort	Esort	Qsort	QsortR	Wsort
2 bytes	812	674	658	703	794	1297
4 bytes	917	894	852	933	1025	1730
6 bytes	1024	1110	1048	1148	1256	2147
8 bytes	1140	1329	1248	1372	1480	2567
10 bytes	1231	1544	1440	1597	1707	2966
MOVE	101020	197504	184675	209963	209780	400637
COMPARE	768789	767910	786726	815920	814311	758977

Table 2. Average execution time to sort 50000 integers (in seconds, 80386DX/40MHz)

Size per item	Fruit [32-2]	Psort	Esort	Qsort	QsortR	Wsort
2 bytes	11.716	9.083	8.968	9.480	11.392	18.762
4 bytes	13.214	12.236	11.679	12.474	14.438	24.460
6 bytes	14.586	15.013	14.226	15.313	17.391	29.838
8 bytes	15.958	17.794	16.782	18.141	20.332	35.245
10 bytes	17.332	20.568	19.313	20.989	23.265	40.640

Table 3. Average execution time to sort n integers (in milliseconds, Celeron/700MHz)

n	Fruit [32-2]	Esort	Esort index	Qsort	Qsort index	Wsort
50000	82.1	46.3	85.7	49.4	94.3	107.0
100000	213.0	99.9	215.9	109.3	246.1	271.8
200000	591.4	225.0	541.5	241.6	628.5	747.4
400000	1543.0	484.8	1274.7	520.1	1439.7	1867.6

References

1. J.W.J. Williams, Algorithm 232,Heapsort 3,Comm.ACM, 7, 1964, pp.347-348.
2. R.W. Floyd, Algorithm 245,Treesort 3,Comm.ACM, 1964, p.701.
3. C.A.R. Hoare, Algorithm 63,64 and 65,Comm.ACM,4(7),1961,pp.321-322.
4. E.H. Friend, Sorting on electronic computers, JACM 3(2),1956, pp.34-168.
5. D.E. Knuth, "The Art of Computer Programming Vol.3: Sorting and Searching", Addison-Wesley, Reading, MA, 1973
6. I. Wegener, The worst case complexity of McDiarmid and Reed's variant of Bottom-Up heapsort is less than nlogn+1.1n, information and computation, 97, 1992, pp.86-96.
7. C.J.H. McDiarmid and B.A. Reed, Building heaps fast, J. Algorithms 10, pp.352-369.
8. S. Carlsson, A variant of heapsort with almost optimal number of comparisons, Inform. Process. Lett. 24, pp.247-250.
9. G.H. Gonnet and J.I. Munro, Heaps on heaps, Proc. 9th ICALP, Aarhus, Denmark, July 12-16,1982,pp.282-291.
10. R.D. Dutton, Weak-heap sort, BIT 33,1993,pp.372-381.
11. J.C. Chen, Proportion split sort, Nordic Journal of Computing 3(1996), pp.271-279.
12. J.C. Chen, Proportion extend sort, SIAM Journal on Computing, Vol.31, No.1, 2001, pp.323-330.
13. A. LaMarca and R.E. Lader, The influence of Caches on the performance of sorting, J. Algorithms 31, 1999, pp. 66-104.
14. R. Sedgewick, Implementing quicksort programs. Communications of the ACM, 21(10), pp.847-857, October, 1978.

Lecture Notes in Computer Science

For information about Vols. 1–2312
please contact your bookseller or Springer-Verlag

Author Index